PALE SKIN, GIANTS,
─── AND THE ───
GREAT TRANSITION

BY MILTON E. BRENER

Copyright © 2022 by Milton E. Brener.

ISBN 978-1-64133-751-9 (softcover)
ISBN 978-1-64133-752-6 (ebook)

All rights reserved. No part of this book may be reproduced or transmitted in any form or by any means, electronic or mechanical, including photocopying, recording, or by any information storage and retrieval system without express written permission from the author, except in the case of brief quotations embodied in critical reviews and certain other noncommercial uses permitted by copyright law.

"The image of the Nordic extratererstrial on the cover is a painting by an artist done at the direction of Michel Zirger, a native of France. Mr Zirger met the subject in Tokyo in 2010. Over the years Mr. Zirger, a journalist and author, had two other encounters with human-looking ETs."

Printed in the United States of America.

Brilliant Books Literary
137 Forest Park Lane Thomasville
North Carolina 27360 USA

OTHER BOOKS BY MILTON BRENER

- *The Garrison Case: A Study in the Abuse of Power* (1969)
- *The Other Side of the Airport: The Private Pilot's World* (1982)
- *Opera Offstage: Passion and politics Behind the Great Operas* (1996)
- *Faces: The Changing Look of Humankind* (2000)
- *Vanishing Points: Three Dimensional Perspective in Art and History* (2004)
- *Richard Wagner and the Jews* (2006)
- *Evolution and Empathy: The Genetic Factor in the Rise of Humanism* (2008)
- *Our Interplanetary Future: A UFO Primer for Skeptics* (2009)
- *Walking Through Walls and Other Impossibilities: The Hybrid Agenda (2011)*

CONTENTS

Preface..9

PART ONE: LOOSE THREADS   11

Chapter I: Transitions..13
Chapter II: Elephants in Our past 27
Chapter III: Ancient Visitors 37
Chapter IV: The Hybrid Agenda 44
Chapter V: Why So Few? 58
Chapter VI: Epigenesis 63
Chapter VII: Pale Skin 69
Chapter VIII: Evolution in Overdrive 75
Chapter IX: Does Evolution Play Favorites? 86
Chapter X: A Case for Recent Change in Caucasian Skin tone........100
Chapter XI: Rule Breakers 109

PART TWO: SO SUDDENLY HUMAN   117

Chapter XII: Miracle at La Marche 119
Chapter XIII: Transition in Africa and Europe 127
Chapter XIV: Of Space and time: The Beginning of

Artistic Genius, and Calendars 134
Chapter XV: Linguistics 144
Chapter XVI: Music... 153
Chapter XVII: One Factor in Transitions 157
Chapter XVIII: The Transition in the Fertile Crescent.................... 159
Chapter XIX: Transitions in Greece and the Renaissance 163

Part Three: The Nordics 171

Chapter XX: Do They Tie the Threads Together? 173
Chapter XXI: Nordics: McClelland and Others................................ 179
Chapter XXII: Steve Wilson 186
Chapter XXIII: Julio Fernandez... 192
Chapter XXIV: Eisenhower and the Nordics 199
Chapter XXV: Worley and His and Other Witnesses........................ 212
Chapter XXVI: Timothy Good and the Case of Albert Coe 221
Chapter XXVII: Blue Eyes and Blond hair 228
Chapter XXVIII: Other Cases of Timothy Good.............................. 235
Chapter XXIX: Ripples from Long Ago 248
Chapter XXX: Musings of a Skeptic 260

Bibliography... 265
Endnotes 271

To my wonderful and wonderfully skeptical wife, *Eileen*,
I am pleased to dedicate this book—even though
she doesn't believe a word of it.

PREFACE

Scientists claim that white, or pale, skin developed because our dark skinned ancestors left Africa about 50,000 years ago and that some went north into Europe. The sunlight, during winter seasons, is less direct far from the equator. Hence, the northern locale resulted in less vitamin D which is manufactured by ultraviolet rays striking our skin. The light skin, which permitted greater absorption, was said to be evolution's response to this deficit. It seemed an ingenious explanation, but seemed to raise many questions. Mainly I wondered about the darker skin of other populations who lived even further from the equator. The explanation of the scientists didn't seem to help.

Scientists also claim there was a remarkable transition about 50,000 years ago involving many new human behaviors and talents. They were, for the first time, so much more like the behaviors and talents of modern people: a better perception of time, and of space, rapid progress in manufacture of new tools, in hunting, and long distance trade, artistic genius, calendric time keeping, and musical instruments. It was thought by most scientists to have been the result of some genetic change in the brain. I could understand that opinion, but I could understand also the claims of some other scientists that it was too rapid for such far reaching changes to have occurred. They believed it must have been more gradual, though there was scant evidence for it, especially for the art, calendars and musical instruments.

I read about writings from ancient times, and saw pictures from prehistory involving strange things in the sky that looked like modern descriptions and photographs of UFOs. The pictures, created many millennia ago, included people wearing outfits with helmets and antenna-looking protrusions. Most, I believed, belonged to a group still very prevalent today, called the Greys. I felt I understood their agenda here, namely to create a race of hybrids to preserve the genes of their dying race.

But I also read about another group of extraterrestrials, seen lately on Earth, who were very different from the others. They usually have blond hair, and blue eyes, like some Europeans. They have pale skin, the only folks around who have skin like the Caucasians, but in their build they look like all of us. They have, sometimes, passed among earthly humans as natives of Earth. They are often extremely tall; some are about 9'. They are the only aliens that consistently make the humans they meet feel at ease, and who arouse friendship rather than repulsion and fear as is sometimes the case with other extraterrestrials. Both military people and the UFO community call them Nordics. But I could not understand the purpose of the continued presence of the Nordics.

I began to feel that these mysteries may be pieces of a puzzle that seemed to possibly fit well together. I turned to the writings of authorities in various fields. I found nothing conclusive, but much circumstantial support for my suspicions and, as best I could, I put it in this book.

Milton Brener,
New York City. November 17, 2012

PART ONE

LOOSE THREADS

CHAPTER I

TRANSITIONS

For some time it has been bandied about in UFO circles that the human race, in its entirety or in part, may be a hybrid. Authors speak of genes of native humans having been combined with those of aliens from another planet. That notion is tied to the discovery of some remarkable technological feats and evidence of sophisticated knowledge in ancient times. Some writers feel these achievements would have been beyond the capability of the contemporary peoples, a belief also fueled by their interpretation of ancient texts and cuneiform tablets. Some of it seems persuasive, but I do not know enough about these claims to make any judgment, and that is not my purpose. I examine instead the possibility of our hybrid origins from a different direction, with different evidence and circumstances.

Some of these circumstances involve proven, or probable, facts of our prehistory. Some have been the subject of debate among authorities with no consensus resulting among them. It involves at least one explanation, namely on the issue of skin color, to which I offer an alternative to that of authorities which does not seem to withstand scrutiny. I started with a search for evidence that hybridization could be the answer to many of these mysteries, and might tie some of them together, or that it could be excluded. The idea is, to me. intriguing, but it poses some knotty questions. When, where, with whom, and over how long a period did such mixing, if mixing there was, take place? Is there evidence, beyond what is offered thus far, upon which conclusions might be based?

This book touches on all of those issues, and ultimately on whether the human race, or part of it, is, or could be, a hybrid. I believe that the evidence supports the view that hybridization is the most probable answer to much that remains unresolved in human prehistory, and which may continue into historical times. It is a scenario which might have been played out beginning in very ancient times.

Archaeologists and geneticists agree that the earliest "anatomically modern humans," (AMH for short), evolved in Africa perhaps as long as 195,000, years before the present, (BP for short, with archaeologists using the year 1950 as the present) and that they were almost certainly dark skinned. The anatomically modern humans are so called because, despite being almost identical to us in anatomy, including brain shape, their behavior and life styles were not very modern. Their brains were, in fact, slightly larger than ours. But their behavioral attributes seemed to have been much closer to earlier human species than to us. The earlier species include populations such as *Homo sapiens neanderthalis* (Neanderthals), *Homo erectus*, and two of the offspring of *erectus* called *Homo ergaster* and *Homo heidelbergensis*. This last named group is believed by some to have been that from which anatomically modern humans evolved. One of the pillars upon which this thesis of hybridization in our past rests, is a gigantic and relatively sudden change in human perception and mentality that occurred beginning about 50,000 years ago.

But it is doubtful, if there truly was hybridization, that it began only about 50,000 YA. Physically our ancestors had resembled the present human race for almost 150,000 years previously, meaning we had also resembled the extraterrestrial race in question, assuming they are in our ancestry, for the same length of time. Even that resemblance could not have come too suddenly, considering the primitive nature of the earlier primates in our lineage. It may well have started with *in Vitro* fertilization before sexual relationship. Such history is almost all speculation, 99% perhaps. Let us look at the other 1%, admittedly not conclusive, but very suspicious.

Dr. Joshua Akey[1] is an Assoc. Prof. of Genome Sciences, Univ. of Washington in Seattle. He and his associates at the University, working to reconstruct modern human evolutionary history, sequenced the entire genomes of five living individuals in each of three different populations. The groups were all African hunter-gatherer populations: Pygmies from Cameroon; the Khoesan-speaking Hadza and the Sandawe, both from Tanzania. The DNA samples had been collected by Dr. Sarah Tishkoff, of the University of Pennsylvania.

The samples were scanned some 60 times, the purpose being to look for genetic code variations that could help explain differences between individuals and populations. One distinctive variant in all three populations studied were genes that produced blood compounds that speeded repair of injuries. The new variants were the same in each of the three groups, though the Pygmies, living in west central Africa, were separated from Tanzania on the southwest coast by over 2100 miles of what was dense jungle for most of history and pre-history.

The Sandawe are an indigenous ethnic group based in central Tanzania in East Africa, and were predominantly hunter-gatherers until the European colonization of the 18th and 19th centuries. Their language, a tonal one with clicks, is apparently related to the Khoe languages of southern Africa.

The Hadza live in north-central Tanzania, in the central Rift Valley and the neighboring Serengeti Plateau. About 300 to 400 of them live as hunter-gatherers, the last full-time ones in Africa. They are not closely related genetically to any other people.

Dr. Akey had no hesitation in terming the particular variants under consideration here as "foreign." In August 2012 some rather startling findings were published. The DNA in each of the five genomes was modern in many respects, but did not resemble DNA from any modern humans or even from Neanderthal DNA. That foreign DNA's influence might explain over time, said Akey, why certain bloodlines of some modern humans heal better than others.

But the most significant mystery may have been the one articulated in *Science Now* of July 26, 2012 "What has most intrigued some researchers is that the study found genetic evidence that all three groups had intermingled sexually with an unknown, older species." In 2012

Akey was interviewed by Linda Moulton Howe, a prominent researcher and author in the subject of extraterrestrials, He stated to her that his group found strong and compelling evidence that in all three of these African populations a relatively small but certainly detectable amount of their DNA comes from more distantly related human groups. But, he added, we don't know which groups they were.

To test their findings they examined non-African individuals for evidence of possible back crossing of hybrids with one of its parent species over many generations, a process known as "introgression." They found that their procedure was correct but it could be ruled out in this case.

The foreign DNA, said Akey, is approximately 1.8 to 2 million years old. As it happens, this was the time when earlier human species were first evolving into Homo erectus, and almost ten times as old as the emergence of anatomically modern humans. Yet the group that contributed it to these African populations, according to Akey, was "much more closely related" to the AMH than to Homo erectus. The only other place where this foreign DNA is found outside of Africa is, not in Europe, but in a handful of places where it is present in the majority of people. He believes that as anatomically modern humans started dispersing out of Africa, they already contained some of that foreign DNA, and it was taken out of Africa with those initial waves of migrations that peopled the world.

Though the population that contributed the foreign DNA is still unknown to Akey, or anyone else, he was predictably dismissive of any suggestion by Howe that "extraterrestrial biological entities manipulated DNA in already evolving primates to create Homo sapiens." She explained to Akey that the quoted words were seen by her in a document shown her twenty years previously at a meeting in Kirtland Air Base in Albuquerque. The document was entitled "Briefing Paper for the President of the United States of America on the Subject of Unidentified Arial Craft."

There was, said Akey, "nothing we have ever seen that would be consistent with that." He described, with a laugh, the claim of ET manipulation as "such a high bar that as a scientist, you have to be skeptical about everything and that just strains the bar of credibility." That was not. he said, "a question that comes to mind when we're trying to interpret patterns of variation. So I'm reluctant to say anything else about it." Experiments to make inferences about history, he continued, should be set up in a way "where we can reject hypotheses or not." He said he would have no idea to know how to begin to design such an experiment.

Asked whether the double helix molecule being in every life form on Earth would support the idea of the seeding of this planet from outside, Akey replied that not every life form has a double stranded DNA. There are RNA viruses and there are single-stranded DNA viruses. He acknowledged that those two were the only cases he knew of. Asked if that implied that the RNA virus could be from somewhere else off planet, he replied "No, I think it implies that we still don't know a lot about the actual origins of life."

In the course of discussion about the unique nature of the DNA in question, Howe stated that it was called foreign because it was completely unknown. That was challenged by Akey:

> It's foreign in the sense that we don't know what group it belongs to, but there is nothing foreign about it. Its DNA sequences like other species have on the planet of the Earth... It's divergent enough that we can spot it in the background of natural variation that exists among different individuals in the world today. But there is nothing extraordinary about it that would demand extraordinary explanation.

Yet he had earlier acknowledged in the interview:

> The group that we think contributed this foreign DNA to these African populations was much more closely related to anatomically modern humans than Homo erectus. I think it is more appropriate to think of this foreign group as about as different as Neanderthals were to Cro-magnon [the name for the earliest fully modern humans] Homo sapiens sapiens.

What is extraordinary is the refusal to consider new and exotic findings because they cannot conceive of an experiment to negate it. The scientific method requires that new findings be repeatable in a way that allows of falsification. The fact that something may be outside the scope of scientific verification leaves many scientists helpless. There have often been new discoveries that were outside the existing scope of standard scientific testing, but later found real nonetheless.

What else is extraordinary is the lack of interest that Dr. Akey shows for truly extraordinary facts. The modern type DNA sequences, with minor exceptions not found outside of Africa, have been found in living persons and determined to be about two million years old. They are closer in kind to that of modern persons than that of Homo erectus, who walked the Earth with them. Yet its importance is peremptorily dismissed.

Let us fast forward, and return to the Earth of about 50,000 years ago.

The change showing basic behavioral attributes that we consider human came about relatively suddenly, whether around 50,000 or, according to others, about 40,000, and to still others around 60,000 BP. It began in northeast Africa according to some archaeologists; eastern equatorial Africa according to others. More than relatively sudden, some aspects of it seem to have come up like thunder. In the words of one scholar the sudden appearance in Europe of artistic genius shortly thereafter was a "creative explosion."[2] Something had happened within the human brain; there was obviously a new way of seeing space and time, and the outside world in general. Whatever the approximate date of its occurrence, it is now marked as the time of transition from the "Middle Paleolithic" to the "Upper Paleolithic," and the birth of a sub-species of Homo sapiens, which we proudly, and perhaps arrogantly, call (ourselves) "Homo sapiens sapiens."

Many scientists of different specialties demur to the claim of such suddenness, and say that that is not the way evolution happens; that it must have been more gradual, and they point to evidence supporting that view. Most of that evidence is based on blade and tool technology, but they can offer nothing on art, music, or use of space on a larger scale than tools or weapons, or calendric time keeping, for which no significant gradualism before about 50,000BP has ever been found.

Innovations almost as significant include houses, built for shelter, some with mammoth bone, but many with dugout floors for semi-subterranean storage, and for hearths and wind protection. There was now, for the first time, sophisticated planning for the hunt, including culling of animals, selection of species by season, mass animal killings, and evidence of hunting forays returning to base camps with meat for storage.[3] The beginnings of communal hunting and extensive fishing occurred at this time, and we see the first conclusive evidence of belief systems centering on magic and the supernatural. For the first time also, as far as we know, sewn clothing was worn.[4]

It is understandable that many cannot accept the vast gulf between ourselves on the one hand, and earlier hominids and every creature now alive on the other. Evolution usually is explained by genetic drift or environmental pressures. Many scenarios involving pressures as underlying this dramatic change have been advanced, but none that are supported by any proof, even circumstantially. It is understandable also that so many cannot accept that such a significant change could have occurred so quickly.

It will be the main thesis here that the entire dramatic change was not through adaptation to the environment or gradual genetic drift, but through another engine that drives evolution, the mixing of populations that had previously lived completely separate from each other. Such a genetic mixture can result in new traits, physical or behavioral, that give nature opportunities to select new, advantageous, genetic combinations. Those changes can spread through a population rather quickly through a process known as epigenesis. Both the effects of population mixtures and epigenesis will be discussed in more detail later. There appears hardly any possibility, and certainly no evidence, of significant mixture of populations in those early times between two earthly populations. But we will later in this essay look outside of Earth for one of the populations and perhaps find a more likely answer.

Once begun, progressively modern behavior continued at a pace more accelerated than the human race had ever experienced. There followed migrations from Africa of a small number of that migrant population into Arabia, western Asia, and Europe, and ultimately throughout the world. Among those who place its beginning in the African equatorial area are Richard Klein and Blake Edward, who specify a site in southwest Kenya, called Twilight Cave (*Enkapune Ya Muto*).[5]

Perforated ostrich eggshells there, dated to 40,000 BP are among the earliest body decorations discovered anywhere.

The migration is estimated by archaeologist Nicholas Wade to have included a mere 150 persons from their base population of about 5000.[6] A study by Dr. Marcus Feldman, a geneticist, however, places the number of immigrants at about 2000.[7] The world population had been drastically reduced about 74,000 BP by a catastrophic volcanic eruption of Mt Toba on the island of Sumatra, and by its lengthy and very toxic aftermath. Whatever the number, and it is agreed to be a small one, every human in the world today, except some in Africa, is descended from these migrants.

Looking at the present world population, such small numbers may be hard to accept. But one underlying source of evidence that can confirm its validity is genetic. To geneticists and anthropologists it is proved by a phenomenon called "founder effect." It occurs when a relatively small group departs from a larger one and establishes a new population entity. This new population will carry only a small part of the genetic diversity that abounded in the old, larger one. Continuing breakup into more numerous populations is sometimes called serial founder effect. The lesser diversity in the new groups results from the fact that each individual leaving the large group carries a much smaller number of the genetic markers than the totality in the larger group they left.

As the new groups expand and age, new markers will occur through mutation for any number of reasons, including pure chance, but they will continue to bear some few markers brought from the original large population. The old population also continues to change, hence the older a population, the greater the diversity of their genetics as compared to the newer ones. It is for precisely this reason that scientists say that Africans are the oldest human populations, namely because they possess the largest diversity of genetic markers of any of the world's populations. All humans carry markers traced to that small population that migrated about 50,000 BP.

In addition, geneticists can estimate the age of separate populations by counting the new mutations, those not carried by earlier groups. Most of these mutations are neutral in effect, but they generally arise at an approximate fixed rate. This, of course, is another important item in reconstructing migrations of the world's populations.

Most scientists say we are not descended directly from any of the more primitive human species, such as Neanderthals, who evolved about half a million years BP, and resided mostly in Europe and the Mid-East. They apparently died out by 30,000 BP, about 10,000 years after the African migrants entered Europe. Nor are we descended, most scientists claim, from other primitive human species such as *Homo erectus* who also seem to have disappeared about 30,000 BP, or *Homo ergaster* or *heidelbergensis*, descendants of erectus. Those groups populated various parts of the world since the first migrations from Africa sometime within the last two million years.[8] Others believe that there was some small genetic input into the new migrants, from the earlier Africans migrants and from Neanderthals in some areas.

Geneticists often speak also of "Haplogroups." These are groupings of individuals with the same genetic characteristic at the same location on the DNA. They tell researchers about migratory patterns and give information about the age of distinctive populations. The haplogroup from which all modern humans outside of Africa are derived is known as L3.[9] It is a "mitochondrial" haplogroup. The mitochondria provide the energy to our cells and are derived through the matrilineal line of inheritance, that is, we inherit our mitochondria solely from our mothers. It has played a pivotal role in tracing the history of the human species. The Y chromosome is carried only by males and is another means of tracing genetic history.

Because the population from whom we are descended is in turn descended from the earlier human species, it may be useful to know something of the dates of the earlier human specimens from whom our immediate ancestral population, and consequently ourselves, are descended. According to author Paul von Ward[10] and authorities upon whom he relies, the existence of "Adam," one of the males from whom we are all descended, lived about 270,000 BP; the original "Eve" from about 250,000 BP. The possible significance of these dates will become apparent later.

It was possibly just shortly after that haplogroup arose in east Africa that a relatively small number of migrants carried it across the Red Sea to Arabia, inaugurating an intercontinental migration that eventually settled every major land mass on Earth except Antarctica. That small group of humans with their later splintering gave rise to several subgroups of the original.[11]

Generally, genetic relationship with Africans tends to decrease with increasing geographical distance.[12] This is one means available to scientists to trace the path of human migrations over the millennia. By such means it has been established that this migration was eastward toward Asia.[13] In following years, groups turned northward through the Fertile Crescent, the land near the shore of the eastern Mediterranean, and another group moved into Europe.

The path to Asia lay across the Red Sea from Africa, in the land now known as Djibouti, and through Arabia. At that time, the world's water levels, including the Red Sea were much lower. This was due to the continuing results of the glaciation, during which much of the world's water was locked up in the glaciers. The lower temperatures during this period may not have affected central Africa a great deal, perhaps 7° or 8° Fahrenheit according to some graphs, but water levels were affected almost everywhere. The Straits of Bab-el-Mandeb, sometimes known as the "Gates of Grief" is the shortest point of the sea between Djibouti and Arabia. It was at the time about 220 feet lower than it is today, and there may have been islands that could have been reached with simple rafts.[14] From there it is believed that the migrants continued to follow the coast, many continuing on to Asia.

Illustration 1: The crossing point from Africa.

Another underlying process that should remove doubts about the smallness of the immigrant group is mathematical. I believe we can assume without special proof, that in that long ago time, sexual activity began in the teens, that there were no means of birth control, and that the subject was not a matter of much concern. Most likely many infants died young, never becoming part of the population, and life was short by today's standards even for those who lived.

Though any estimate of world population of that early time would be mere guess work, those who guess seem to congregate at about 250,000. How do we get from 150 or 2000 to the seven billion descendants of today? The following numbers from Vaughn's "World Population Growth history,"[15] may be of interest. They deal with the years from 10,000 B.C., (12,000 BP), the earliest date for which that source gives an estimate, to the year 2011. The population reached 1,000,000 (all figures are, of course, approximations) in 10,000 B.C.; 5,000,000 in 5,000 B.C., or a fivefold increase in five thousand years. In the next 5,000 years, at the time of Christ, it had grown to forty times as much, to 200,000,000. Less than two thousand years later, in 1830, the population reached 1,000,000,000, namely one billion, a fivefold increase (despite the crippling effects of the "black death" plague). In the next hundred years, by 1930, it doubled, to 2,000,000,000. By the year 2000 it was 6 billion, 80 million, more than a threefold increase in 70 years. In 2011 it was 7 billion.

This new phase of human culture, beginning about 50,000 years ago, is called in Africa the "Late Stone Age," (LSA). In Europe and elsewhere, it is called the "Upper Paleolithic," (UP), meaning the latest of the three recognized stages of the Paleolithic, which itself means "Old Stone Age." The population expanded rapidly, thanks in part to selection of new game and new methods of hunting. Klein and Edgar place the migration as spreading into western Asia about 45 to 43,000 years ago; into eastern Europe between 40 and 38,000; and into western Europe between 38 and 37,000 BP.[16]

Various cultures within the western European Upper Paleolithic are identified by particular tool types and other distinctive items.[17] They each lasted for a span of about 5 to 8,000 years

The *Aurignacian* Culture was the earliest. It stretched from Bulgaria to Spain between about 37,000 and 29,000. There followed the *Gravettian* Culture, which extended from Portugal across southern and central Europe to European Russia, There then arose the *Solutrean* which existed in France and Spain, and the last, the *Magdelenian*, which included France, northern Spain, Switzerland, Germany, Belgium and southern Britain between about 16, 500 and 11,000 BP.

There is apparently no evidence from the extensive archaeological research of warfare between them, though some dates must have overlapped. Nor is there any indication that any of them disappeared as a result of conflict. This could raise serious questions as to whether cultures that didn't slaughter each other really could be our ancestors, and questions also as to how they spent their spare time. With regard to the second mystery, it appears they frittered away much of it painting pictures of animals on cave walls. Never mind that they are among the finest animal paintings in all of history and far excel any in prehistory. They also engaged in other frivolities such as playing flutes they carved from animal bones, probably the first music to be heard by humans.

With regard to the first question, some archaeologists, interestingly, have trouble believing, regardless of the lack of evidence that these cultures did not engage in combat. Wade starts with the proposition that "A propensity for warfare is prominent among the suite of behaviors that people and chimpanzees have inherited from their joint ancestors."[18] It is indeed embarrassing that the chimps, having long ago engaged in the art of mutual slaughter, the skill was never developed by the humans of the UP. Wade insists that since "warfare was an incessant preoccupation of early human existence," that the picture of the UP of western Europe, drawn by the specialists is "strangely incomplete." The specialists should really get busy on that.

Wade is not alone in his discomfort. In a volume entitled *The Origins of War: Violence in Prehistory*[19] the authors acknowledge that though it is tempting to dismiss those populations as "fraternal, calm, societies, feasting on the bountiful fruits of nature," lack of archeological evidence makes such assumptions difficult to prove or to disprove. The authors prefer to focus primarily upon the most advanced stages of prehistoric society: the Neolithic and the Bronze Age, where evidence of warfare is so much more visible.

But another authority[20] seeks to soothe our disappointment by assuring us that evidence for warfare in one UP area is available. From 14,000-10,000 BP we see the use of new weapons such as bows, slings, daggers and maces in organized warfare, depicted in cave paintings, mostly in Spain. The art of western European otherwise created during the Upper Paleolithic however reveals nothing so progressive. The occupants were obviously too busy painting animals and playing the flute.

We turn to something more serious. The earliest of those dates of humans in western Europe mentioned above, about 40,000 years ago, is about 5000 years before the date of the earliest paintings in Chauvet Cave in south central France. Cosquer Cave, whose entrance is now under the waters of the French Mediterranean, is just a few thousand years younger. Those caves contain the oldest known art of that artistic culture. The first population to enter Europe was termed the Cro-Magnon. After about 10,000 years they had occupied much of that continent. They were the progenitors of the later cultures, just mentioned, that dominated Europe and created the splendid sculptures, and paintings on cave walls and ceilings, for about the next 26,000 years.

We will later look more closely at this transition, and the remarkable changes in human perception and behavior that it included. The next significant transition came much later, only about 11 to 13,000 BP, with the advent of farming and settlement of the first villages and cities. It was centered in the lands known as the Fertile Crescent, those bordering, or near, the eastern Mediterranean shores, primarily in what is today Lebanon, Israel, Jordan. Syria, Kuwait, Iraq, northwestern Egypt, and extending into Anatolia, modern day Turkey. The new period it ushered in is called the Neolithic, or "New Stone Age," not to be confused with the Upper Paleolithic, which is the most recent stage of the Paleolithic, or "Old Stone Age."

We will also take a look at two transitions in historical times involving important changes in human thought, feeling and behavior, one in ancient Greece and Rome, the other in the western European Renaissance, beginning about the 14th or 15th Century. The possible genetic underpinnings for those fluorescences may help us to understand the causes for the earlier ones in prehistoric times which are of primary interest to us here. Circumstances indicate that those later creative periods were sparked by the mixtures of populations who had for many millennia lived entirely separately from each other. They will be discussed only for that purpose.

CHAPTER II

ELEPHANTS IN OUR PAST

There is another change that occurred no earlier than the remarkable developments beginning at Twilight Cave. According to a majority of geneticists and evolutionary biologists, it was between 20, 000 and 50,000 years ago that there occurred the first appearance of pale skin on humans, sometimes wrongly called white, which admittedly sounds better. It developed in the wave of immigrants into Europe in that period. According to the accepted theory, light skin was said to be evolution's response to the less direct sunlight of northern areas, permitting as light skin does, greater penetration of ultra violet rays. We will in Chapters VII and VIII take a much closer look at that proposition.

According to another authoritative source, the first appearance of pale skin came much more recently. It developed at about 12,000 BP according to one source, and at about 5500 according to still another. Both dates happen to fall within the period of the second, and much later, prehistoric transition mentioned, from hunting and gathering to farming and permanent settlements. These geneticists also believe that the light skin made its appearance in the Fertile Crescent at that time, for which they attribute the change from meat and fish diet, rich in vitamin D, to one of grains which contain practically none, a matter also to be discussed in Chapters VII and VIII. It should be said at the outset however, that the evidence does not seem to support a change in diet so much as a supplement to it.

It is also worth pondering whether there was any pressure guiding the evolution of pale skin in the Mid-East which is mostly in the sub-tropics. As articulated by one authority. Dr. Michael Holick, "If you live in the tropics or subtropics, you have access to UV-B rays from the sun year-round, so it's quite possible to get most or your entire vitamin D requirement from responsible exposure to sunlight."[21] The tropics are the area of the Earth between 23° north and 23° south. He defined the sub tropics as between 23° and 35° in latitude, north or south, from the equator. We will later hear more from him and more about those numbers.

We might ponder: if the residents of the Fertile Crescent had adequate sunlight for Vitamin D purposes, why did they ultimately develop pale skin like their European neighbors to the north? Whatever mixing they had with Europeans, and, as we will see later in Chapter X, it apparently was not much, it resulted from immigration from the Crescent to Europe, not the reverse.

What are UV-B rays? While the solar ultraviolet spectrum is continuous, scientists find it convenient to describe it within three specific wavebands—UVA, UVB and UVC—classified according to their wavelength, UVA being the longest. They differ in their biological activity and the depth to which they penetrate into the skin. UVA, the longest wavelength, accounts for up to 95 % of the solar UV radiation reaching the Earth's surface. It can penetrate into the deeper layers of the skin, and play a major part in skin aging and wrinkling, and hence may also initiate and exacerbate the development of skin cancers. UVC is the shortest and contains the highest energy. However, it is filtered by the ozone layer, as long as we still have it, and consequently does not reach the earth's surface and does not contribute to skin damage in humans.

The stratospheric ozone layer also absorbs most of the harmful ultraviolet UVB radiation emitted by the sun. In small amounts, UVB radiation is helpful to life. It is UVB that is involved in the production of vitamin D. But it is now been discovered that UVB, as well as UVA rays can produce DNA damage in the skin and in mucous membranes. This can cause genetic mutations that can lead to skin cancer, premature skin aging, cataracts and other eye conditions. Excessive UV exposure can also weaken immune system functioning, reducing our ability to fight off skin cancers and other maladies.

UVA penetrates more deeply into the skin than UVB. It also causes more genetic damage than UVB in skin cells where most skin cancers arise, namely in the basal, the innermost, layer of the epidermis. UVB tends to cause damage in the more superficial epidermal layers. It is responsible for burning, tanning, acceleration of skin aging, and also plays a very key role in the development of skin cancer. The intensity of UVB varies by season, location and time of day. The most significant amount of UVB hits the U.S. between 10 AM and 4 PM between April and October. UVB rays do not penetrate glass or clouds.

UVA rays, on the contrary, are present during all daylight hours and throughout the winter months. Although UVA rays are less intense than UVB rays, they are present all year round and depending upon the time of the year, can be 30 to 50 times more prevalent than UVB rays. Furthermore, UVA radiation can penetrate glass and clouds. Thus, we are exposed to large doses of UVA throughout our lifetime.

Exposure to the combination of UVB and UVA is a powerful attack on the skin. It can create irreversible damage that ranges from sunburn to premature aging to skin cancer.

The purpose of melanin is to absorb UV radiation and dissipate the energy as harmless heat, blocking the UV from damaging skin tissue. UVA gives a quick tan that lasts for days by oxidizing melanin that was already present, and it triggers the release of the melanin from melanocytes. However, because this process does not increase the total amount of melanin, a UVA-produced tan is largely cosmetic and does not protect against either sun burn or UVB-produced DNA damage or cancer.[22]

By contrast, UVB yields a slower tan that requires roughly two days to develop, because the mechanism of UVB tanning is to stimulate the body to produce more melanin. However, the initiation of production of melanin by UVB, called melanogenesis, requires direct DNA damage by it. Hence the cost of the protection by dark skin may be damage to the DNA, the building blocks of the body.

So it appears that nature may have played a dirty trick on us. What happens to us in advanced age may be of no significance to evolution, but it might be of great significance to us, especially with the lengthening life expectancies. So much for intelligent design or for a benevolent evolution.

Vitamin D, in reasonable amounts is very beneficial, especially in advanced age, in fighting cancer. But the UV rays that cause the synthesis of vitamin D plays a large hand in causing cancer. Proof was obtained by the Wellcome Trust Sanger Institute, in Hinxton, UK. The coauthor of the report. Professor Mike Stratton, MD, PhD, explained that the mutagens occurred years before the tumor became apparent. "We can see the desperate attempts of our genome to defend itself against the damage from ultraviolet radiation. Our cells fight back furiously to repair the damage, but frequently lose that fight. According to the Sanger Institute, "The melanoma genome contains more than 33,000 mutations, many of which bear the imprint of the most common cause of melanoma—exposure to ultraviolet light."

It's a rough world. It truly looks like you can't win, but we obviously aren't programmed to win. We're programmed to lose. It is almost as though a repair man came into your home to add protections of some kind, but turned out to be a clumsy oaf who might have caused as much or more damage than he prevented. Much of the above has been discovered within the last decade and has apparently not yet seeped into the consciousness of the public.

Before our detour we were talking about the mid-East and the Fertile Crescent. Apropos of the latitudes that define the sub-tropics, the areas between 23 and 35° north or south, the locations of the present countries that comprised the Fertile Crescent invite scrutiny. They seem to show access to sufficient sunlight rendering diet less crucial. All references here are to latitudes north of the equator: Iraq, which encompasses the most important part of the Crescent, is almost entirely below the 33rd; Kuwait is almost entirely south of the 30th; Syria is geographically cut in half by the 35th; Damascus being on about the 33rd; Jordan is entirely below the 33rd; Baghdad being on that parallel; Israel, including the West Bank is entirely south of the 33rd; the southern border, about the 31st; Turkey is between the 36th and 46th; Egypt is between the 20th and 32nd.

In short, the availability of sunlight in the Fertile Crescent, with the possible exception of Turkey, seems to undercut the whole idea of evolution of skin color driven by lack of sunlight. I do not doubt that the immigrants in Neolithic times had light skin, but neither diet nor inadequate sunlight seems to have been the driving forces. The driving force, of light skin in the Fertile Crescent, whenever the change came, and whether before or after the change in northern Europe, may well have been whatever it was that caused the light skin in Europe. Mixture with light skinned extraterrestrials would suffice.

Though scientist argue about when and where light skin first appeared, I, as a layman, see little reason why the two entries in the dispute cannot both be correct. In that case the European side would have been earlier, but that would not rule out a separate genetic change, whether caused by the same or different genes, in the Fertile Crescent thousands of years thereafter. There is evidence that it probably did not occur through any mixing of the two populations.

That evidence comes from the work of geneticists. There were those who believed that the genes of the earlier immigrants into Europe, largely hunters and gatherers, now predominate, and others who believed that the predominant genes now are those of the later entrants, the Neolithic farmers. That issue will have greater relevance in response to an argument to be examined chapter X and will be discussed more fully at that time. We will state at this time only that the evidence supports the predominance of genes of the early hunter gatherers, the spread of farming in Europe apparently resulting from acculturation, learning, rather than genetic predominance of new populations.[23]

There seems to be agreement that by 30,000 BP the Cro-Magnons had occupied most all of Western Europe. Even the Russian Arctic, by some evidence. was settled by 40,000 BP. There is, in any event, only one significant migration out of the east by the first farmers across central Europe in late prehistoric times, namely the one mentioned above. It occurred about 7500 BP over a period of about 500 years. Though farming began in the Fertile Crescent about 10 to 12,000 years ago, the migration of farming people, called the Linear Pottery Culture, or the "bandkamerik" came from Hungary and Germany, to which the farming occupation had spread. That latest date is shortly before the Bronze Age.

The beginning of written language occurred in Sumer, now Iraq, in the Fertile Crescent about 4500 BP.

It was the descendants of the group that left Africa about 50,000 BP that replaced the older archaic human groups, and undoubtedly brought with them the L3 Haplogroup. The genome of the smaller group entering Europe about 37,000 years ago no doubt had, or soon developed, the gene (SLC24A5for those who are interested), found in the genome of almost all Europeans and of those of European descent. It is in those populations that ubiquitous pale skin abounds. It is difficult to see how such an abrupt change in skin tone in any later times, such as the 10,000 BP, or 7500 BP, could have so quickly spread throughout Europe if the earlier settlers in that continent from 40,000 BP, had gotten along so well for thousands of years without it.

We should, at this point, deal with another matter of this nature, one involving the East Asians. It has been often said in the literature that genetic mutations lead to light skin in both East Asians and Europeans,[24] though involving different genes. It indicates, write some authors, that the two groups experienced a similar change due to settlement in northern latitudes.[25] Such parallel phenomena have been termed "convergent evolution." A group of scientists investigated the evolution of pigmentation variation by testing for the presence of six genes that affect skin tones. It was found that forms of 2 genes may play a shared role in shaping light and dark pigmentation across the globe, and were said to be probably a major cause of the light skin of East Asians. We will see in Chapter VII that three other genes play a predominant role in the evolution of light skin in Europeans but not in East Asians.[26]

To speak of lighter pigmentation is fair enough, no pun intended. Often however authors speak of the white skin of West Eurasians (including some relatively small part of west Asia), or the white skin of both East Asians and Eurasians, thus equating one with the other. The skin of the Caucasians in the west is not white; it is pale, truly without color, or light pink, caused by a myriad of small veins close to the skin.[27] The East Asia skin is not white either. It is light tan, bronze or copper toned. Scientist measure darkness or lightness of skin by a "reflexivity index." The process by which it is measured need not concern us. It is enough to know that the scale runs from 20, very, very dark, to slightly

PALE SKIN, GIANTS, AND THE GREAT TRANSITION

over a hundred, for very, very light. Natural human skin color diversity has been found to be highest in Sub-Saharan African populations,[28] with skin reflectance values ranging from 19 to 46 (median, 31). This compares with 62 to 69 for Eurasians and 50 to 59 for East Asian populations.[29]

The Europeans thus have a reflexivity for which the smallest number is 3points higher than the highest number for the East Asians. The highest number for the Eurasians is 10 points higher than the high point for the East Asians. How does the geography fit into this scenario? The lands involved as East Asia are vast, including western and central Russia, Japan and China, among other areas. Russia is northernmost, stretching from 45° in the south to 75° in the north, which is far more north than the British Isles. China stretches from 15° to slightly above 45°. Japan runs from 30 to slightly higher than 45. Hence the populations of these lands despite comparatively light skin and the high latitude are darker than those of the Fertile Crescent or even of those of northern Europe, except Lappland.

The following illustration 2 is part of a study dealing with skin tone of various world populations. The young lady shown was illustrative of the skin tone of East Asians, though whether Japanese. Chinese, Russian or otherwise we are not told. Nor does it matter. She is obviously a great advertisement for whatever racial heritage she may have.

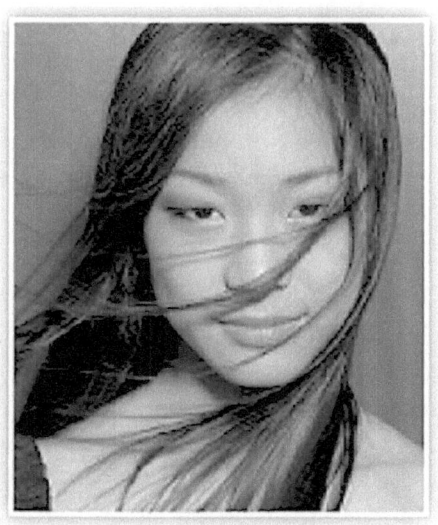

Illustration 2: The light tan of the Far East.

Ill 3, below, shows the variation in color of the various populations, and, if not so eloquent as the above photo, makes the point better than all the words of the geneticists. Only three genes are shared between East Asians and North Europeans, and they do not include SLC24A5. That gene is said by some geneticists to account for about one third of the pale tone to the skin of North Europeans.

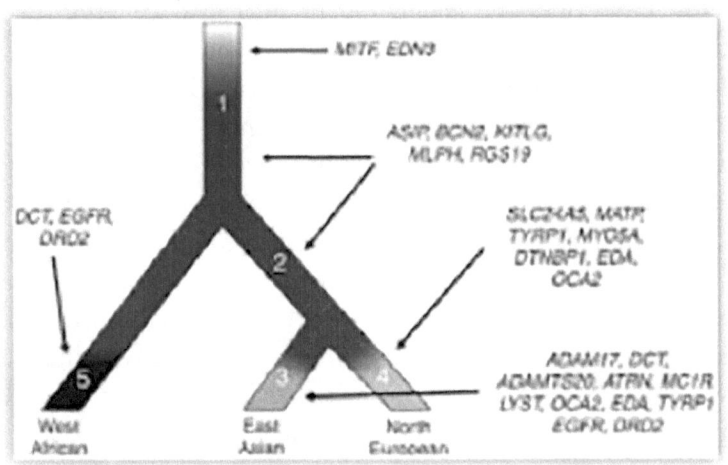

Illustration 3: The evolutionary genetic architecture of skin pigmentation in three populations. Despite gene flow among human populations, differences in the allele frequencies at pigmentation genes are still observed. The genes listed have demonstrated signatures of selection as (1) shared among all humans, (2) shared between East Asian and European populations, (3) unique to East Asian populations, ((4) unique to European populations, and (5) unique to West African populations.

Illustration credit: http://www.gnxp.com/blog/2006/09/another-genetics-of-skin-color-review.php

Before we get too bogged down in this, we should take cognizance of the plain fact that skin color means nothing with respect to any individual. Human beings can be judged only by their qualities as individuals, their accomplishments, and their characters as shown by their behaviors, not by skin color. Identifying skin color of persons, or fossils, or long gone populations can have probative value for limited purposes, such as tracing the history of migrations in the past or for other purely historical purpose, or matching a person or remains with reported descriptions, but not for making character judgments. There is in short a scientific purpose at times that makes such an inquiry legitimate. To make it for purposes of judgment about behavior or character is a fool's errand and entirely pointless. It is a most unfortunate thing that I should feel it necessary to state such a truism, but I am aware that racists still walk the Earth, and I do not want to furnish fuel for their absurd fires.

It all leaves us with still another question that can, at this juncture be answered only with guess work. The question: Why us? Not us Earthlings; that is not hard at all. Earth may not be unique, but its proliferation of life is undoubtedly something unusual and highly attractive to any exploring or wandering extraterrestrial group. But why do only the Europeans and their descendants have the ETs' skin, if indeed we do? If the circumstances do in fact add up, and if the sudden surge in human genius occurred with their assistance, why did only those humans who migrated to Europe wind up with such pale skin, skin that resembles none other on Earth, except the Nordic ETs?

The skin tone of each of the other originally geographically separate populations is unique, but they have this in common: they all have color. The colors they have may have indeed been developed through lack of sunlight and sufficient animal sources in the diet, but they all could be seen as remnants of black or brown. But how and why are Caucasians colorless, or for those who prefer, lily-white?

I dislike creating scenarios without evidence. But sometimes it seems obligatory.

We can perhaps envision that the group. 150, or 2000, or however many there were, who struck out for new horizons, proliferated, grew in numbers over 10,000 years or so and that different groups went different ways. Perhaps also one subgroup came, more than the others, to resemble the Nordics in skin tone and facial features. That would not be unlikely, but very probable by the rules of genetic inheritance. It can be seen on a smaller scale in our own family histories and genealogies. Some children resemble their parent in some aspect or another more than others. Some seem to get preference, consciously or otherwise, for that reason.

Perhaps the Nordics may have continued to visit, and to show most interest in, and possibly most romantic attraction to, the lighter skin and other similarities to themselves of this particular group. Assuming all of those perhapses, that particular migrant group would have been more and more endowed with the genes, and the coloring, as well as the features, of the cosmic visitors. It seems to be a possible scenario and one perhaps more likely than the unlikely explanations contrived by scientists to explain the pale skin.

CHAPTER III

ANCIENT VISITORS

WE NEED TO LOOK AT still another matter of interest. To a certainty there is much convincing evidence of the presence of extraterrestrial aliens and of UFOs on Earth going back untold thousands of years, no matter what say the skeptics. The term, UFO (unidentified flying objects), has taken on a secondary meaning, referring not to craft that are merely unidentified, but which, because of their exotic wingless structure, silent flight, and remarkable maneuverability, are deemed beyond the capability of any human technology on Earth. It is in that sense that the term will be used here.

Proof of the longevity of their presence on our planet consists chiefly of artworks, medieval, ancient, and prehistoric, and credible reports of strange sightings that go back over two thousand years. One of the very oldest of the specimens of that art can be found on the walls of several of the caves in southern France and northern Spain. I have considered one of the more convincing to be a line drawing on the wall of Pech Merle Cave, in southeast France, dating to about 15,000 years ago (ill 4).

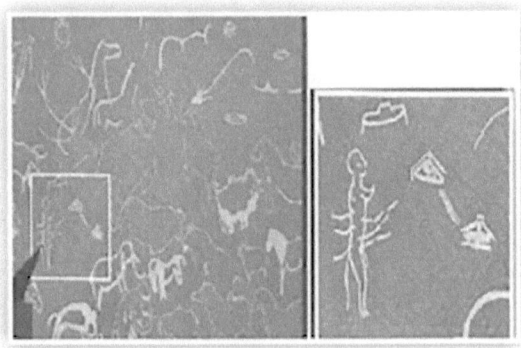

Illustration 4: The photograph on the right is an enlargement of the content of the box outlined in white on the photograph of part of the wall of Pech Merle on the left.

Credit: Matthew Hurley. *The Alien Chronicles*, Quester Publications

The top drawing in the blowup in Ill 4 is similar to many modern photographs of UFOs, exemplified by the illustration 5 shown below. Illustration 6 below is the photo of a model for an aquatic UFO built by Paul A. LaViolette, and appears similar the two lower engravings shown in illustration 4 above.

Illustration 5: March 19th 1993 about 6:00 A.M. Looking west from Ochre Point Cliff, near Adelaide, on the southern coast of Australia, from about 400 meters. The outline appears similar to the top engraving in the blowup (right photo) of ill 4.

Credit: www.ufo.se/maslinbeach

Illustration 6: front view of UFO, as explained in the text above: to be compared with the lower two figures in ill 4, the blowup on the right side.

Credit: Paul LaViolette, Secrets of Antigravity Propulsion, 285, and P. Potter.

Another, of the same approximate age, and almost as similar to modern photographs of UFOs, is in a number of the caves in the vicinity of Ussat (ill 7) in south central France. The other caves include Cougnac, Niaux, Font-de-Gaume, and Pair-non-Pair caves in France, and Altamira, La Pasiega, and El Castillo caves in Spain.[30] Most have been dated to between 12,000 and 18,000 years ago. A particularly intriguing one involves the one from a cave near Ussat, in the center of the group. It shows what could be antenna on the sides and a ladder on one side from the ground to the main structure. In many details the entire image is remarkably like the photograph in Illustration5, taken in 1963 in Italy. That is especially true of the antenna like protrusions, and for the existence of what appears to be a ladder or stairs that was most probably not in the repertoire of our ancestors in Paleolithic times.

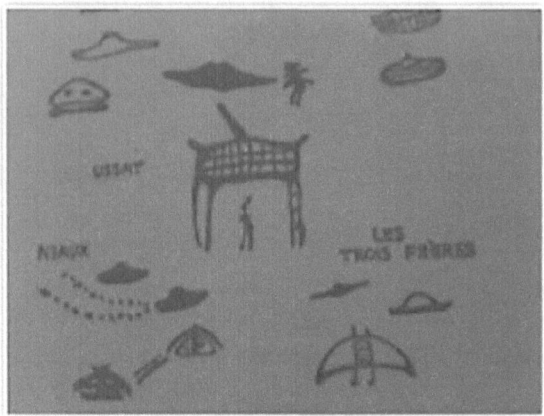

Illustration 7: The two engravings in the lower left are similar to the bottom two in the right hand photo of ill 4. Many of the others appear similar to the disks described and photographed in recent times. The central figure is the most interesting, for reasons described above. It has been suggested that the figure beneath, whether human or humanoid, is to show comparative height. It may be more likely to merely depict an occupant of the structure.

Credit: Ralph and Judy Blum: *Beyond Earth,* p. 40.

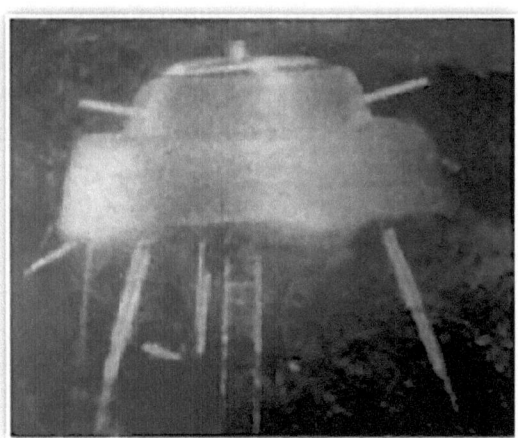

Illustration 8: According to Timothy Good, this photo of a landed UFO in Genoa was submitted anonymously to an Italian magazine, *Domenica del Corriere* in June 1963. The photographer claimed to have seen it with a garage worker, but that for personal security reasons would not give his name.

Credit: Timothy Good and *Dominica del Corriere*

One of the most convincing bits of such evidence, though more recent, is a painting from the 15th Century, presently on exhibit in the Palazzo Veccio in Florence, Italy. It is known as The Virgin and Saint Giovannino, and in part shows a man gazing upward at an object that looks remarkably like many of the UFO depictions and photographs in modern times (ill 9).

Illustration 9: As can be seen from the composite photographs above, the right hand part of the composite is a blowup of the detail showing a man with his dog staring at a flying object. The insert is a further blowup showing more detail of the object which is quite similar to very many other objects seen, described and photographed today, including the radiation that is sometimes observed.
Credit: Palazzo Veccchio, Quester Publications, and Matthew Hurley.

There are also paintings on rocks in the Americas and elsewhere of strange human like creatures who looks like they wear helmets very similar to those worn by our astronauts, sometimes with antenna looking objects projecting. Some are dated as early at 13,000 years ago.

In addition there are thousands of written reports of sightings of strange "ships" or other craft, or of unexplainable lights in the sky from very early times, extending back to several centuries before Christ. Some could be explainable phenomena. Others could not. Many accounts are from Patrick Cooke and his extensive research as compiled in his *UFOs in History*,[31] from *The Alien Chronicles*, by Matthew Hurley, and from *Wonders in The Sky*, by Jacques Vallee and Chris Aubeck. We see many recurring themes, extending, like threads, right into recent times. In the more recent ones, later than about 1800, many sightings are by reputable astronomers.

Many of the reports speak of flying shields, one of the earliest coming from Alexander the Great in 329 B.C. Many early shields looked like triangles, a word commonly used today to describe many of the mysterious flying objects. There are other frequent descriptions of these strange objects matching those of the present day, such as spheres (or globes) and disks. Still another similarity between ancient and modern sightings is the phenomenon of the "mother ships," which are large ships that discharge and receive smaller ones. The "'Nuremburg Chronicles" describes a strange fiery sphere, seen in 1034, soaring through the sky in a straight course from south to east and then veering toward the setting sun. The illustration accompanying the account shows in the air a cigar-shaped form. Cigar shapes are frequent descriptions of large UFOs in modern times.

In Japan, in 679 it was reported that cotton-like matter, about 5 to 6 feet long fell over Nariwa (now Osaka), and was "drifted by the wind here and there." Something called "angel hair," in recent decades often witnessed in several countries, including the United States, has been described as long white stuff, slowly descending from UFOs, and exhibiting some very peculiar properties. There have been only partially successful scientific investigations of it nature.

Gervase of Tilbury wrote that in 1207 in Tilbury, England, an aerial ship caught its anchor in a pile of stones. An occupant came down from the ship and managed to free it, however he was asphyxiated by the atmosphere. General George C. Marshall, in 1951, reported that on three occasions landings had been disastrous for UFO occupants. He stated that breathing Earth's heavily oxygenated atmosphere had burned them to a crisp.[32] A flying large silver disk was seen by many monks and abbots in northern Yorkshire, causing "utmost terror." It fits the description of objects reported in recent times, including Finland in the current century.

In Japan in 1235, while General Yoritsume and his army were encamped, mysterious lights were observed in the sky winging, circling, and moving in loops. The general ordered a "full-scale scientific investigation." Reported the investigator: "The whole thing is completely natural, general. It is… only the wind making the stars sway." Some modern explanations from "authorities," are hardly any more credible.

Omitted from this summary are hundreds of reports from scientists, mostly astronomers, from the last two hundred years of mysterious lights on the moon. They actually begin in the second half of the 18th Century, though some are as early as before Galileo.

So what is going on? Whatever it is, it has been going on for a long time. I doubt that it started in the 1950s, or even in 329 B.C. It would not be inappropriate to quote at this juncture from Dr. Edward S. Gilfillan, Jr.'s book. Migration to the Stars, published in 1975:

> *Judging from our experience here many planets should have developed technological cultures by now, and if we have a good chance of colonizing planets so had they. Some of them should have made it long before us, say a billion years earlier... Why then haven't we seen them?*[33]

Maybe we have. And I can sooner believe that they came a billion years ago than that they have been on the scene a mere 50,000.

CHAPTER IV

THE HYBRID AGENDA

Obviously, before humans could write, or were creative enough to paint or draw, there could hardly be any way of knowing aliens were here, unless they purposely, or otherwise, left some proof of their presence. This, even in more recent times, they try not to do. Painting and engraving came well before writing, by at least 30,000 years. If aliens are in fact in our ancestry, who might they have been, and why are they here?

There are perhaps many different alien races, probably from many different places. As to the precise number, that must, for the time being, remain for the eye of the beholder. Between some groups there are very few, and often very minor, differences. Between others the differences are strikingly apparent. Analogous to this, anthropologists and archaeologists divide themselves into "lumpers" and "splitters" in their categorization of human fossil bones on Earth. The lumpers will find relatively few species of ancient humans, as they believe that many differences will occur in the same species. The splitters are more apt to find many species, often naming new ones based on relatively minor differences. Further complicating the whole subject is the near certainty that some descriptions by not so credible witnesses are most probably figments of the imagination. But not all of them can be so easily dismissed.

If earthly scientists cannot agree on the number of different species in human evolution, we could hardly expect agreement in the categorization of extraterrestrials. But is there truly any reason to think we are a race of hybrids with any of them?

We will here be interested primarily in one such race of extraterrestrials, termed originally by our military personnel, as "Nordics." The overriding thing about them is that, far more than any other E.T. group, they look like us. If they are in our genealogy, and if their genes are entirely or partly responsible for the abrupt change in human behavior that occurred about 50,000 years ago, that episode was probably not the last one involving input from their genome into ours, and almost certainly not the first one. Much earlier must have come sufficient mixing for the physical similarity between them and ourselves. We have seen some possible evidence for that in Chapter I. For though their skin color may be similar to that of only some of us, their body build is like that of all of us, except for a few relatively minor differences, and except also perhaps for the height of the very tall ones.

What other evidence is there for the earliest input from the genome of the alien visitors into ours? Perhaps it is a matter for speculation only, but we should note the treatment of this question by Zecharia Sitchin in his highly popular *The 12th Planet.* His conclusion comes primarily from the ancient texts, most importantly cuneiforms from ancient Sumer and from the Bible. He translates and interprets the texts to reveal, in summary, that the early astronauts visited Earth about 445,000 BP; that one of their activities here was the mining for gold, for very practical manufacturing purposes. Sitchin states that the aliens were termed *Nephilim*, meaning, he says in that volume, the "people who fell down upon the Earth from the heavens,"[34] despite the fact that in common usage the word means giants. In his later volume, *There Were Giants Upon the Earth*,[35] he retracted his denial of the common meaning and through some rather complex logic and interpretations acknowledged that the Nephilim could indeed mean giants. To the ancient Earthlings these were gods, and were so treated.

The rank and file of these aliens, "those who performed the tasks," termed Anunnaki,[36] were put to work by their superiors, continues Sitchin, mining the gold and doing other physical labor, such as smelting, mostly in Mesopotamia. About 300,000 BP these Anunnaki refused to work any longer and the nephilim, probably through *in vitro* fertilization, created armies of workers by hybridization between the *Homo erectus* natives of Earth and themselves. This was necessary to produce creatures that were physically and mentally able to perform the labor for which they were created, but still mentally inferior to themselves, the Nephilim.[37] He finds the evidence in ancient texts including, among other places, the ancient *Epic of Gilgamesh*. In this connection we should recall the dates of the first "Adam" and "Eve," 270,000 and 250,000 BP respectively according to modern geneticists.

Thus, says Sitchin, The Nephilim "took an existing creature and manipulated it, to 'bind upon it' [quoting the ancient script] the image of the gods." Further, writes Sitchin, "Man is the product of evolution, but modern Man, *Homo sapiens* is the product of the 'gods,' "actually the astronauts from what Sitchin terms the 12th Planet. Hence, about 300,000 years ago, the Nephilim took ape-man (*Homo erectus*) and "implanted on him their own image and likeness... without the creativity of the Nephilim. modern Man would still be millions years away on the evolutionary tree."[38]

The first efforts at hybridization, again from Sitchin who interprets his sources, did not go well. The first creatures were very imperfect, a woman who could not bear children, a being who had neither male nor female organs, a man with diseased eyes, sick liver, failing heart, trembling hands, a man unable to hold his urine, or sicknesses "attendant upon old age." It all sounds like some of the monstrous results of early hybridization efforts by our scientists with earth born animals. But finally, came perfection, one with the "skin as with the skin of a god" smooth and hairless, quite different from the hairy ape-man." It was with this final product that the Nephilim found so attractive and were able to many and have children with them[39]

This attraction and resultant marriages are placed by Sitchin about 100,000 BP. This will be mentioned further, in Chapter XXVII in more detail including the account in the Bible. Assuming some validity to

Sitchin's account, the occurrences of 100,000 BP may mark the beginning of anatomically modern humans. If so, the transition to *Homo sapiens sapiens* occurred about 50,000 years later. That origin of anatomically modern humans however is firmly believed by modern scientists to have occurred about 195,000 to 200,000 BP.[40]

When the Nephilim first sought marriage with the women of Earth is described in the Bible in Genesis 6: 1-4, but, like almost everything in that sacred volume, it gives no hint of the time frame. The 100,000 BP date deduced by Sitchin is too late, by scientific estimates for the beginning of anatomically humans; too early for the transition to the Upper Paleolithic. It could however be right on target for the first marriage between the Nephilim and human women, there being no reason to conclude that the dramatic change in mentality happened so quickly following the earliest interbreeding. It is solely for this reason that the Sitchin narrative is summarized here. One hundred thousand BP, more or less, could be the beginning of hybridization resulting in *sapiens sapiens*, or it could be nothing at all.

There is no reason to believe that the interbreeding between humans and Nordics from space, if it in fact occurred, ceased 50 or 20 thousand years ago, or even 5.5 thousand years ago. We will later deal with the case of the Spaniard abductee, Julio Fernandez, another of a former US military officer, and another involving a high ranking employee of NASA, among many others, which seem to be evidence that their presence on Earth continues to the present time.

They are in most cases, though in not all, indeed tall. Some are apparently however no taller than the average human. Their heights seem to vary reportedly between 5'6" and 9 feet in height. They are pale-skinned, with colorless lips, blond haired when they have hair, and blue eyed. They possess the same facial features as humans, though usually markedly sharper as described in detail by one of their abductees. Like a number of other alien groups, some of them have eyes, said by some, though not all witnesses, to have pupils that are oval and vertical, seeming like slits, rather than the circles formed by human pupils.

They are generally considered to be the "good guys" sometimes referred to, by the more romantically inclined, as our "space brothers." From descriptions and histories they could almost be said to take a parental interest in us. Their presence on our planet within very recent times has been attested to by many including a number of very highly credible witnesses.

The Nordics not only look like us in body type, but, most often, they react well with humans, and humans with them. Many of them can, and have, passed for human. Probably the same discrepancy in height existed 50,000 years ago. Today, six footers or even six and a half footers are no rarity.

Seven feet will cause a startled second look from most strangers. Eight, and nine feet in height are almost unknown, the few specimens usually attracting news coverage. Whatever the height of these aliens, there is often reported positive feelings evoked in their human contacts.

They are apparently the most numerous of the encounters in the United Kingdom. "They have been described there as the most human-like of extraterrestrials. Portrayed as having long blond, almost platinum hair and amazing blue eyes, they are very beautiful with the males having muscular physiques."[41] The same source reports that they seem to be the ETs that most often interact with human beings and to be interested in social issues affecting humanity. Two women who bore their children are named. So was British Air Marshal, Sir Peter Horsley who, in a well-known and well reported episode, claims he was visited by an alien who warned him about nuclear weapons. From the description it was probably a Nordic.

According to the *Chambers Dictionary of the Unexplained*,[42] they have been described as benevolent or even "magical" beings who want to observe and communicate with humans. Those who have had contact with them have said that the Nordics are concerned about the earth's environment or prospects for world peace, and may transmit messages telepathically.[43] An American social worker[44] has said that they are described by those he interviewed, as "paternal, watchful, smiling, affectionate, youthful, [and] all-knowing."

Stephanie Kelley-Romano, a highly respected associate professor in theater and rhetoric at Bates College describes the Nordics as associated with spiritual growth and love and often act as protectors for humans they have contacted.[45] A few "contactees" say the Nordics have warned them about the Grey aliens,[46] a few examples of which we will see later in this book. Others say they have seen Nordics inside the same craft as greys.[47] In such reports, the Nordics are often interpreted as leaders, with the greys as their subordinates.[48] Jenny Randles, a British author and former director of investigations with the British UFO Research Association, writes that although she believes Nordics have "certainly" been involved in abductions she feels the abduction is "less essential to the encounter than it is with the [greys]."[49] According to UFO investigator, Michael Salla,

> There is an "apparent pattern that has occurred in Ufology in the dominance of the space brothers in the 1950's who were kind, interacted with people who became known as contactees, and took people for rides in their space crafts. This pattern changed dramatically with the abduction of Betty and Barney Hill in the early 1960's. The space brother human types of the 1950's seemed to fade away, and they were replaced in the UFO literature with another type of alien. In the early sixties the first abduction of the Hills began a new pattern where the aliens were grey "evil" aliens who would abduct people against their will and perform medical procedures on them. There were, as far as this author is aware no confirmed cases of "classic" abductions in the 1950's. Unlike the "good" space brothers of the 1950's these Grey aliens were described by all, who were unfortunate enough to have met with them, as being distant and without emotions."[50]

That last sentence is largely, but not now entirely correct. Emotions are suppressed in the Grey, but some witnesses claim to have seen signs that it still exists. Unfortunately, we know far less about the first group, obviously Nordics, than about the Greys. Though they probably do not figure into our limited scope of inquiry, we should talk a bit about the Greys before going further. They and their agenda and methodology are an important part of the entire picture of aliens on Earth. It may shed some likelihood of hybridization in the distant past by another race, namely the Nordics.

Beginning with the public awareness of the UFO phenomenon, about 1947, the Greys have far outnumbered other groups of extraterrestrials, at least in the North American continent. The Greys are those frail little creatures, 3 to 5 feet tall, with the large heads and the huge black eyes, with small slits for a mouth, holes for nostrils, and with no, or little, protrusion of a nose. Their presence on Earth has been attested to by too many, some of whom have too much credibility to ignore. They are called the Greys because of the greyish tincture to their skin. They are the ones we instinctively envision when we hear about extraterrestrials. It is they who have been the subject of countless drawings and caricatures. They have apparently won the admiration, or sympathy, of some of our fellow Earthlings, but only hatred and fear from many others.

Illustration 10: Drawing of a Grey.
Credit: http://exopoliticsjomnal.com/vol-3/vol-3-2-LoBouflo.htm

An English source reports[51] that of all extraterrestrial encounters, the Greys are the most famous. They account for 43% of encounters in the USA and 90% in Canada. In Europe however they account for only 20% and in Great Britain 12%.

The *modus operando* of these Greys includes, almost without exception, the extraction of semen from the men; ova, that is, eggs, from the women. These are apparently ultimately combined with reproductive material of the aliens, *in vitro*, thus creating hybrid offspring. *In vitro* fertilization involves placing eggs and sperm together in a laboratory environment, then transferring the fertilized egg into a woman's uterus, or womb, where implantation and embryonic development will hopefully occur just as in a normal pregnancy. It seems, from descriptions, that for the Greys, sexual intercourse may be a thing of the past.

There is also a clear possibility of creation of "transgenics," a human being, conceived by human women, with an egg fertilized by human semen. Even when carried to term by a human mother, if the egg or embryo has been injected with one or more of the alien genes, either in the womb or *in vitro*, it becomes a transgenic. If a physical or behavioral trait is caused by the alien gene, the transgenic's offspring and future generations may inherit the trait. The researches of a number of our scientists, experimenting with animals, have established the truth of these statements beyond doubt to any with open minds.[52] A monkey egg or embryo, for instance, injected in such manner before birth with one particular gene from a jellyfish, has caused the monkey after birth to glow a bright green.[53]

Why do the Greys abduct humans, and what are they up to? The evidence adduced from study of the entirety points to the fact that the Greys are probably a dying race. Our own scientists tell us that other planets likely to have intelligent life are apt to be, on average, 1.8 billion years older than Earth. This means two things of possible significance for us. First, any visitors from other planets are likely to be much further advanced technologically than are we, and secondly, their planets are nearer their lifes' ends. But for their advanced scientific and technological achievements, it appears that the Greys, at least, may have paid a very high price. Their race seems to be in serious decline. Their emotional life has atrophied, and there is little of it left.

What is their intention with regard to the Earth? Apparently they look upon our remarkably fecund planet, so rich in the necessary ingredients of life, as the ideal home for hybrids. It would be a race consisting of a combination of their genes with their advanced scientifically oriented lives, including perhaps advanced intelligence, and ourselves with our strong emotional makeup, most particularly the bond between the mother and her children. It appears that their own children are conceived by means of *in vitro* fertilization, and the embryos grown in incubators until "birth," at which time they still lack the strong emotional closeness to a mother that normally begins in the womb. The aliens seek not only hybridization, but also, as a necessary adjunct, learning to behave as humans. Many, perhaps most of the abductees on Earth, have been taken by them repeatedly and the women subjected to various gynological procedures, including removal or implanting of eggs or embryos. In addition they also seek from the women, demonstrations of bonding, including breast feeding and cuddling.

Between humans and the Greys, there is much tension. True, there have been many contacts between abducted humans and Greys that have gone relatively smoothly. Some of our people, mostly women, have looked with renewed spirituality and kindly feelings about their experiences. Conception with them is often, though not always, a laboratory thing, done *in vitro*. It appears that the Greys have lost not only most emotion, or at least the expression of it, but the ability for normal sexual relationship. Some women have interacted well with their abductors, and bonded with the offspring with as much love and affection as they would with an entirely human infant. Others refuse all insistence that they raise the child born of such a union, many with reactions of extreme anger reflecting hatred of the aliens and even of the infant.

One of those abducted, a trained psychotherapist, said about her captors' society that it seemed to be dying and that "children were being born and living to a certain age, perhaps preadolescence, and then dying… It is a culture without touching, feeling, nurturing… Whatever their bodies are now, they have evolved from something else. My impression is that they had wanted somehow to share their history and achievements and their present difficulties in survival."[54]

How can there be hybrids, or even transgenics between Earthlings and natives of another planet when, according to the dictates of modern biology, we are obviously not of the same species. We deal first with the hybrids.[55]

It should be emphasized that all of the hybrids created by our scientists to date are between species of the same "order." There are, in short, no offspring spawned by a combination of a lion and a camel. Our main point of interest here is that some alien groups, particularly the Nordics, seem closer to earthly humans in anatomy and physiology than many of the combinations we will see.

Among the hybrids that exist, notwithstanding the impossibilities articulated by many biologists, are ligers, which are crossbreeds between a male lion and female tiger, and tigons, which are crossbreds between a male tiger and a female lion. Ligers are the world's largest cats.

Then we have the complex situation of the ti-ligers/ti-tigon/li-tigons/li-ligers. They constitute a crossbreed between a male tiger and a female liger/tigon or a male lion with a female tigon/liger. Female ligers or tigons are fertile. In the case of ti-ligers. they have unusual striping where it breaks up and displays a blotchy appearance. Since they are 3/4 tiger, they exhibit more characteristics of a tiger than a lion.

A leopon is the result of breeding a male leopard and a female lion. The head of the animal is similar to that of a lion while the rest of the body resembles the leopard. Leopons are larger than leopards and, unlike lions, like to climb and swim. Crossbreeds between dogs and wolves are fairly frequent, and the dog-wolf hybrid will display a wolf behavior or dog behavior or something in between. Domestic Tamworth pigs are sometimes crossbred with wild boar to create "Iron Age Pigs." The hybrids are tamer than wild boar, but less tractable than domestic swine.

A wholphin is a rare hybrid, a cross between a bottlenose dolphin and a false killer whale. A wholphin's size, color, and shape are intermediate between the parent species. The first captive wholphin showed mixed heritage even in its teeth: bottlenose dolphins have eighty-eight, false killer whales have forty-four and the wholphin with sixty-six has split the difference. A zebroid is a cross between a zebra and any other equid such as a horse. Because of the huge differences in sizes of animals which sometimes make natural breeding impossible, they are often created by artificial insemination.

In short, though it is still in the elementary biology books that species cannot breed with each other, they can. Look, if you will, at the strange mixtures of characteristics displayed by these hybrids, and see if the similarities and differences between us and descriptions of our many visitors are any more pronounced.

We turn to the subject of transgenics.[56] It has been said that breakthroughs in molecular biology are happening at an unprecedented rate. Included is the ability to engineer transgenic animals, animals, that is, whose genome has been changed to carry genes from other species. The technology has already produced transgenic animals such as mice, rats, rabbits, pigs, sheep, and cows, though not necessarily in that order, either numerically or chronologically. The mice, however, are presently, both chronologically and numerically, in first place.

All of that has been done by our own scientists. Imagine, if you can, what might be accomplished in that field by a race with even a few thousand years advantage over us.

But why would they want to come to our planet to perpetuate their genes? The more cogent question: Why not? Earth, in all likelihood not unique in the universe or even in our own galaxy, is almost to a certainty something very unusual. Our astronomers have discovered over 750 exoplanets, planets that revolve around other stars, and they tell us that none seen so far seem conducive to life. The ubiquity of life on Earth need not be belabored here, though the extent of it is probably not commonly known: more than two million species of life known and named, and probably at least ten million more still unknown. In every cubic meter of our indoor air, there are up to ten million cells of bacteria, some of which are of great benefit to us.[57]

It bears repeating that the behavior of the Greys is worth mentioning, as much of the same methodology and motivation may, to a probably lesser extent, apply to the Nordics, or may once have done so.

Perhaps one of the aliens who said it best was a Nordic. He was talking (in mental telepathy according to the abductee) to a Spanish man, resident of Madrid, captured on a road northeast of that city and taken into a UFO. The man, Julio Fernandez, about whom we will hear more later, told investigators that he had never believed in flying saucers and that seeing humanoids approach his stalled car was like Karl Marx beholding God. My opinion that those aliens seem to have been members of the Nordic race of ETs is based largely on drawings made by Fernandez (ill 45 and 46, Chapter XXIII) and by his descriptions.

Significantly, according to Fernandez, he was told that another race was coming to Earth, that it was not as "ethically evolved" as themselves, and that this other group was involved in "probing and programming the minds of humans."[58] That could only have referred to the Greys. That is precisely what the Greys do, in ways that are difficult for Earthlings, and for some, impossible to believe. It is important to note, as we will again later, that his captors took a sample of semen from Fernandez.[59]

But there is little point in dwelling further on the Greys here. We leave them in order to concentrate on the Nordics, which in the last part of this tome we will do. Despite having much less knowledge about the Nordics, than about the Greys, it is the Nordics with whom we will here be primarily concerned.

Assuming the length of the history of visitations by aliens, what does it mean in relationship to our past? There has been much speculation in recent years centered on the remarkable feats of early civilizations on Earth in their erection of monumental structures, including the ingenuity required, the skillful design, the precision in the shaping of the materials used in construction, and the movement, both horizontal and vertical, of materials of tremendous weight. The proponents of this view cite as examples the Egyptian pyramids, and other impressive ancient structures throughout the world. The arguments favoring alien presence and intervention center also around the suddenness of the development of such skills, claiming that without advanced outside help we would expect to find a much more gradual growth of it. Knowledge of mathematics and astronomy have also been found to be far more advanced than could be expected without outside help. The surge in technical know-how, and scientific knowledge, they claim, was too sophisticated to have been accomplished without help which was not available on Earth.

The behavior of the aliens, however, as far as we can tell from present evidence, has been one of nonintervention, except in the breeding and development of hybrids. That would not be definite proof that in earlier days they, or other alien races, would not have been more involved in earthly matters. But if our current visitors are any indication, it appears more likely that the mixing of our genes with theirs would be the extent of their mission, or at least of their behavior, except for study and analysis of our human race. The intellect of the hybrids, if the present behavior of the aliens is any guide, was to be trusted to enable them to make their own progress.

According to one episode from 1954, such hands-off attitude exhibited by the Greys was also the policy of the Nordics. Air Marshal Sir Peter Horsley has already been mentioned in these pages. He was former Deputy Commander-in-Chief of the British Strike Command and had flown 90 different types of aircraft beginning in WWII. He spent seven years in the service of the Queen and Prince Philip as Equerry. It was during that latter period that he had an experience of interest to us, one of profound effect on him.

According to his autobiography,[60] he was not convinced that UFOs were from another planet. A friend, a General Martin, on the other hand, was convinced that they were being contacted by extraterrestrials seeking to warn us of the dangers of nuclear warfare. One day in 1954 Horsley was introduced to a Mr. Janus. The various contacts through which the meeting was arranged, though interesting, are not of great moment to us. The meeting involved only Horsley and Mr. Janus. Present also was a Mrs. Markham who was instrumental in setting up the meeting but took no part in it, serving merely as observer. According to Horsley, Janus seemed to fit quite perfectly into his surroundings, normal in every way, except, writes Horsley, he gradually seemed to take over the conversation. "My initial reaction was one of scepticism," writes Horsley, "but by the end of the meeting, I was quite disturbed, really." The meeting lasted two hours, but what is germane to our subject was short. It was however very relevant to perennial questions from skeptics.

Mr. Janus said that "The basic principle of responsible space exploration is that you do not interfere with the natural development and order of life in the universe any more than you should upset or destroy an ant heap or beehive… The observers [space travelers] are not interested in interfering in your affairs, but once you are ready to escape from your own solar system it is of paramount importance that you have learnt your responsibilities for the preservation of life everywhere."

He had earlier explained that the space travelers must "insure that evidence of their existence is kept away from the vast majority of Earth's population. You must be well aware of the damage which your own explorers have done by appearing and living among simple tribes… only the most developed societies can cope with such contact."

Horsley made a full report to Prince Philip. Whatever else he was, said Horsley, "Janus left me with the impression of a force to be reckoned with. He appeared to know a great deal and spoke with authority about space technology." Prince Philip's private secretary thought it was a joke. But neither the Prince nor Horsley were so sure. They both retained open minds.[61] About Mrs. Markham we know nothing. She appeared out of nowhere, and seemed, after the meeting to disappear into nowhere.

There is one other item that may, or may not, be circumstantially corroborative. Art Campbell, a veteran of the U.S. navy during the Korean War in the field of electronics is also a retired educator and author of four books on northwest pioneer history and UFO crash investigations. He is also a researcher into matters involving UFOs and extraterrestrials generally. He quotes correspondence from a contact in the United Kingdom who had access to the super-secret M15 files: "In the 1953-55 timeline, the ET visitors had landed at several places and asked for a meeting with the leader of the most powerful country on Earth." Though there is no account in the report of Sir Peter of Mr. Janus requesting such opinion, the report of the British contact is included for what anyone wishes to make of it, if anything. Meetings between Eisenhower and ETs in that timeline will be dealt with in Chapter XXIV.

CHAPTER V

WHY SO FEW?

It is beyond question that sunlight enhances production of vitamin D and that dark skin in the higher latitudes of our planet, that is, those more distant from the equator, whether north or south, can inhibit the body's production of that vitamin. These do so, primarily, because of the slant angle of the sun to the Earth at those latitudes. It results in more of the sun's ultraviolet rays at the higher latitudes being weakened or lost before contacting Earth's surface, and the skin of its inhabitants. There is also the factor that in winter the days are shorter.

Neither is there any dispute about the fact that deficiency in vitamin D can result in rickets in infants, particularly in the first two years after birth. There seems also adequate evidence that osteoporosis, a disease that usually strikes the elderly, 60 years of age or over, seems to correlate with infantile rickets.

What is open for debate, and by some authorities is debated, is the question of whether the claim that the pale skin on Europeans or those of European descent was caused solely and entirely by genetic changes in response to the migration of the dark skinned humans into Europe beginning about 40,000 years ago. This involves other questions as to how long ago it and any other causative factors occurred and where the occurrences were. They may be matters that can at this time be neither proved nor disproved definitively. Many circumstances however seem, to point to between 50,000 and 20,000 years ago, either in Africa or in southern Europe.

Several important aspects of evolution should be at least mentioned in passing at this point. First, phased bluntly, evolution does not care about older people. The number of older people dying has no effect on continuation of the population, the sole "concern" of evolution. One of the two major adverse consequences of lack of vitamin D, osteoporosis, for instance, does usually not strike until advanced age, well after the great majority of its victims have produced all of the offspring they are likely to produce. The subject of rickets, which affects only the very young, may be another matter, but as we shall later see, not necessarily.

Regardless of the cause however, it is the point of this essay that the change in latitude, moving north, is most probably not responsible for all of the change in skin tone of the group known today as Caucasians. Whether it has caused some skin tone changes in Europeans and other populations is not denied, but is not crucial to our main thesis. We call our skin white, but what it is, is colorless. White is the blank page of typing paper, the color of clouds if not saturated, milk, a banana, our teeth, the skin of a polar bear. That is not our skin color. We are pale. The Native Americans had it right: palefaces, or so they used to say in the movies.

Caucasians are the only native population on this planet without color. Others are black, brown, tan, red, or yellow, bronze or copper toned or something in between. The extraterrestrials are sometimes called, by Caucasian witnesses, as pale or white. No group native to Earth is pale except for Europeans or their descendants. And none are really white. Why did evolution single out Caucasians for this alleged honor? If we have a pinkish glow it is due to the myriad of tiny blood vessels just below the skin. Only through our colorless skin do they make much of a difference. As the map, illustration 11, shows, the lightest skin is European and only European. The numbers are not at all correlated with the reflectance scale mentioned earlier where the higher numbers represent lighter skin rather than darker skin. In illustration 11 below, the numbers are exactly the opposite.

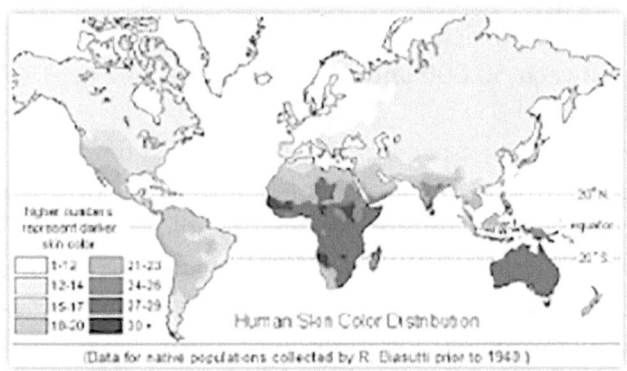

Illustration 11: Global map of skin pigmentation levels

The proposition that our pale or white tone, call it what you will, is a result of the loving care bestowed on us by the tender mercies of evolution without outside mixing is highly suspect. It contradicts much that is told to us by geneticists about the evolutionary process. Evolution is not a nursemaid to the human race or any part of it. It is not a nanny charged with the task of protecting us against all dangers of nature or, certainly not of relieving us of discomfort or suffering. About such things, evolution could hardly care less. It "cares" only about survival.

Look, if you will, at the multitude of diseases to which "the flesh is heir to," to quote another writer, and at their continuation, and about which evolution is oblivious. Look at the much greater multitude of causes for them that are still with us, about which evolution doesn't seem to be bothered. But in short order, we are told, evolution saw to it that we got our share of Vitamin D, no matter how relatively trivial the result of some shortage may be in survival of the species.

We should realize also that evolution, in the traditional understanding, which seems to be hanging on with many scientists today, does not actively "do" anything. It is a passive force. Those that cannot adapt, that cannot compete, or cannot survive in a rather ruthless world, simply cannot reproduce. As long as they can reproduce sufficiently to perpetuate the species, then the species survives. If not, it ultimately, usually, soon dies out. As long as the species can reproduce at a satisfactory rate to survive, evolution sits itself out.

Another aspect of evolution involves the subject of "epigenesis," for some reason still ignored by most writers on the subject. It will be discussed in detail in the next chapter. It undoubtedly played an important role in the change of skin color following the migration. There will be some things about the change to the pale skin of Caucasians however that neither the standard theory, nor epigenesis can explain.

With that in mind let us go back about 45 to 50,000 years. The amount and degree of change in human behavior, over a few thousand years, in human creativity and intellectual accomplishment equals, or perhaps eclipses, that occurring during any previous or recent epoch. That growth spurt is probably one of the most significant and most dramatic behavioral changes in human history. This "transition" and its causes are believed by many scientists to be genetic in nature, some change, that is, within the internal structure of the human brain and nervous system, but the matter is still debated and a matter of much scientific controversy. The scientists have no fossil brains surviving from that early time to compare with ours of the current time, only the size and shape based on fossil skulls.

No matter when our skin turned in tone so drastically, there seems no dispute about significant changes in behavior and skill changing in the same time frame, whether 50, or 10 thousand years ago. Nor is there any dispute but that it was a relatively small group of one culture that left Africa about 50,000 BP. Those facts in themselves raise some interesting questions.

Why did such a small group leave their home and their fellow residents of the place called Twilight Cave? What was different about this small group? Perhaps it was curiosity, or a sense of adventure, for instance, that overcame the prevalent fear of leaving familiar territory. But it was probably something more basic, and more vital. This was during the latest period of glaciation which began about 110,000 BP, ending about 12, 500. and reaching its period of maximum glaciation about 20 to 18 thousand BP. Conditions in east Africa, though not glaciated, were cooler than usual and arid. The aridity probably meant lowered availability of vegetation including fruits and vegetables "gathered" even if not cultivated, and possibly game. Food may have been scarce. If it was hunger, was this small subset of the group somehow disadvantaged in obtaining foods, or were they simply more resourceful than their fellows. Or were they driven out by the others for some reason?

Is it possible that the pale skin did not occur entirely as a result of migrations to the north, but that instead, they were lighter skinned than the others of their population before leaving Africa? If so, two possibilities come to mind.

First, lighter skin may have made it uncomfortable for them in or near equatorial Africa. Temperatures today in Kenya reach a high during the day in the 90s, all year round, 95 and up in January, February and October, other months a degree or two less. The drop in temperature at the height of the last Ice Age worldwide was about 8° Fahrenheit, though as much as 20° in some places including much of Europe. Even 8° was not much alleviation for Kenya's usual heat. It was also a period of aridity, and the skin of the migrants may have already been turning lighter.

Perhaps the climate or scarcity of food, or both resulted in their migration, being continually drawn toward the cooler, or colder, climate. It may be the most acceptable explanation of the migration, which possibly reached the Russian Arctic nearly 40,000 years ago.[62]

The other possibility is that prejudice against them may have played a role. They may have been forced to leave. Humans, many of them, have long made it known that they do not like anyone who is different, especially of different color. It can make some of us, even in these "enlightened" times, fearful and angry. How much more so in those long ago times, we can only guess.

It is possible that the Nordics may have been around longer than we presently speculate, and it may be that the migrants of that first transition, or later ones, or several of them, were hybrids. Interbreeding with humans may have been a continuing thing for many millennia before 50,000 years ago, as discussed in Chapter IV. But we will look at first things first.

There are problems with the standard explanation for pale, skin. We will look at the basics of the theory, and then at some of the problems with it. But we should first look at something that should be borne in mind, if only as a reason for not getting involved in the question of whether there has been time enough for such changes to occur.

CHAPTER VI

EPIGENESIS

It must be acknowledged that the time required, even for the late date, for the change in skin tone is not really a factor. What removes the time element is the phenomenon of epigenesis. What follows here is not intended to advance any conjecture I make about the spread of light skin, and it indeed does not do so. It may in truth provide ammunition to those espousing the relatively rapid spread of it, such as those, for instance who believe the appearance of pale skin may have occurred as recently as 10 or 12 thousand, or even 55 hundred years ago. But for whatever it proves or does not prove, or whatever proposition it tends to support or weaken, it is part of the total picture, and should not be ignored.

It is nonetheless remarkable, though not unusual, how many mainstream scientists have ignored the recent proofs, in both nature and in the laboratory, of very rapid evolution that cannot be ascribed to pure chance mutation of genes.

Until fairly recent times, it was taught and preached that most evolutionary change occurred as a result of accidental "mistakes" in the production of eggs by the female or sperm by the male. All of the cells of the body, other than those engaged in reproduction are continually dividing into two daughter cells which are genetically identical to the parent cell. In the case of the sex cells however, the egg and the sperm, there are four daughter cells, each containing one half of the chromosomes of the cell, as were present in the parent cell. When this happens genetic diversity is increased because the chromosomes in this process undergo a recombination which shuffles the genes producing a different genetic combination in each cell.

It was part of the standard theory that most results of mutations in the sex cells, the only ones that mattered in evolution, were neutral, harmless, but of no benefit. Those that did have an effect were usually harmful ones. But on rare occasion, the results caused some advantageous change in the environment in which the individual lived, and those changes spread throughout the population and possibly beyond. Whatever the cause of change, it was a slow and gradual process, except for what has been called "punctuated equilibrium," a sudden spurt amidst lengthy slow processes. No explanation of the cause for such bursts of rapid change has ever been put forth. It was simply descriptive of what was observed.

Under standard theory there was no direct feedback to the genes from the environment or from the living habits of the individual. Nor did it matter what happened to the genes other than those involved in sexual reproduction.

There are three billion "base pair," or units, in the human genome, meaning an individual's entire complement of DNA. A very tiny percentage are genes, 25,000 approximately, each consisting of anywhere from a few to a few thousand base pair, or units. The remaining units were called "junk DNA" by biologists, as they hadn't a clue as to what use they could be. Genes code for proteins, but the other units, consisting of most of the genome, do not. Many considered this huge percentage to be left over from evolution during a time in which they played some role. This is another case of a "just so" explanation, taught as gospel by researchers who really had no idea as to the significance of the non-coding units, which they called "junk DNA."

In recent years, however there has been much scholarly activity concerning new understandings of the process termed "epigenesis." It has been defined as the "interface between the social and biological levels of organization,"[63] and involves most of the human genome that geneticists formerly called junk. It was eventually discovered that much of it is not junk, but plays a vital role in evolution due to its relationship to the world outside of the genome, including the physical and mental behavior of the organism of which it is a part. Unlike the protein coding genes, it responds to outside stimulus through the stimulus's effect on human behavior.[64] New behaviors, in short can result in changes, both physical and behavioral.

For one thing, it may make instinctive various behaviors that have been learned by earlier generations through trial and error, sometimes called "the Baldwin Effect" in honor of David Baldwin who first articulated the process. According to Daniel Dennett, it explains how "A practice that is both learnable (with effort) and highly advantageous, once learned can become more and more easily learned, [and] can move gradually into the status of not needing to be learned at all."[65]

Charles Lumsden and Edward O. Wilson were pioneers in this as in other areas of biological research. In 1981, they wrote that substantial evolution of cognitive traits in a thousand years or less is very likely to have proceeded right into modern times.[66] They refer to "co-evolution," meaning mutual feedback between environment and the individual's genome. It is not limited to cognitive traits, but that is, for us, a very important part of the process. They argue also that the rate of change depends on the amount of variation in the epigenetic structure. Selectivity, in turn, they say, "is affected by the amount of heterogeneity [variation in genes] in the environment." Such variation can flow from intermarriage, particularly between populations who have lived completely separate and have had separate genetic histories.

The epigenes play a significant role, for another thing, in determining when a gene will be active, and when it will be quiescent. It can also, cause change similar to that of genes, but usually on a relatively temporary basis. It can even cause evolutionary type changes in the organism for a number of generations. Through a process called "canalization," it can create a path for the next gene, new or mutated, and thereby make it more likely that it will be expressed in a certain way, and temporary traits thus made permanent. It has in these functions taken out, or greatly reduced, the element of blind chance in human evolution.

According to Nessa Carey in her recently published *The Epigenetics Evolution*:

> The epigenetic system controls how the genes in DNA are used, in some cases for hundreds of cell division cycles, and the effects are inherited from when cells divide. Epigenetic modifications to the essential blueprint exist over and above the code, on top of it, and program cells for decades.[67]

It appears that genes have many possible manners of expressing themselves, not merely one or two as had for so long been supposed, and the paths taken may often be those set out by the epigenetic pathways. Conrad Waddington was the first to use the term epigenesis, in 1942, and also first to speak of canalization in this context.[68]

Each of the many occurrences of epigenetic change were carefully and scientifically observed and verified and would, by many scientists, probably not otherwise have been believed, and would possibly have been ridiculed. I cite only three. I see little need to dwell on a subject that apparently may have little bearing on ours, but I feel that some explanation should be made of the proposition that evolution of traits, when adaptive, can happen very quickly. It does not have to wait for a chance mutation. The three examples, in very short form, follow:

In 1981, certain guppies were taken from a river filled with their predators and relocated upstream to where only one of the predators lived. In eleven years, they had grown bigger, delayed sexual maturity, and lived longer, even in relation to those not eaten by predators in the earlier environment. This change was found to be up to ten million times faster than the average rates determined from fossils.[69]

In 1997 Anolis lizards in one of the Bahaman Islands were transported to various other islands in that chain, each having vegetation with limbs of varying thickness, but all being of smaller diameter than those on the island from which they came. Over a period from ten to fourteen years, representing seven to ten generations, the hind limbs of the lizards decreased, each in proportion to the lesser thickness of the limbs on the new island.[70]

An episode in Holland gives evidence that such epigenetic changes occur in humans, and that they are not always beneficial. They seem to follow the habits of one or more generations. During World War II there were severe restrictions in the diets of Dutch women imposed by the Nazi occupiers. This caused a large range of disorders in themselves and their children, and extended in many cases to following generations despite diets of adequate nutrition. Physicians found it remarkable that a pregnant mother's diet could affect her descendants, possibly for an untold number of generations, despite a lack of change in the genes.[71]

Nessa Carey was not completely convinced that this was a case of the epigenetic mechanism. She pointed to other possibilities such as an abnormality in the cytoplasm of the egg that could stimulate an unusual growth pattern in the fetus that could in turn result in a transgenerational effect. This would not be through inheritance of an epigenetic modification. Perhaps, she reasoned, transgenerational effect from fathers to children, if found, would be more certain evidence of an epigenetic mechanism.

To address this issue she cited a study of mortality patterns for male descendants of men who were alive during a certain period in the early 20th century involving severe food shortages interrupted by periods of plenty.[72] Of particular interest was the food intake of male descendants of men who lived through a well-recognized childhood "slow growth period" (sgp) during those episodes of shortage versus plenty. The scientists deduced that if the food was scarce during a father's sgp the sons were at decreased risk of dying through cardiovascular disease. If the father had access to a surfeit of food during the sgp, he had an increased risk of dying as a consequence of diabetes.

Carey felt it reasonable to hypothesize that the transgenerational consequences of food availability were mediated through epigenetics. She emphasized that the original nutritional effect had happened before the offspring had even begun to produce sperm.

These are but a few of many hundreds, possibly thousands, of such occurrences that have been observed and studied in recent times. They all involve initial changes, bodily or behavioral, without changes in the genes. Any change in the genetics came later if at all. For very many decades, despite the approved dogma that genetic changes were the result only of accidents in cellular copying of genes involving reproduction, and matters of pure chance selected by nature for their advantageous traits, laymen and scientists alike obviously never really internalized it.

They always seemed to speak of evolutionary change as though it were anything but blind.

In any event, as we talk about evolution of skin color, it will not hurt to keep these matters in mind, even if only in the back of it. There seems no necessity for evolutionary change of this magnitude either genetic or epigenetic. The change, in the long run may have been more deleterious than advantageous, one more thing to be discussed later.

CHAPTER VII

PALE SKIN

THE ULTIMATE TEST OF THE validity of this theory of the evolutionary cause for our pale skin, or any other scientific theory is whether or not it "works," that is, whether the same facts in other areas produce the same results. There are a number of areas on the planet where the theory can be tested. These would the ones where humans have lived for many thousands of years removed from the equator by the same or greater distances as those inhabited by Caucasians. We should look to see if evolution has shown in such cases results similar to those found with Caucasians in Europe, namely pale, colorless skin

One of the major proponents of the late change of skin tone of the Caucasians, namely, 12,000 BP or later, believes that a combination of lack of direct sunlight and reliance on cereals and grains as opposed to animal based foods was the cause. There are at least two major instances. Native Americans and Far Easterners, where both conditions were met, but did not result in the development of pale skin. First however we might respond to the majority of scientists who see lack of direct sunlight as the sole cause of change of skin tone, and see its beginning 20 to 50,000 BP in Europe. And before starting to answer that viewpoint, we should look at the basis for it to begin with.

The condition most often mentioned as the trigger for the evolutionary step in skin color is the disease known as rickets. It involves the softening and weakening of bones in young children, often because of an extreme and prolonged vitamin D deficiency. It is a condition not often seen in newborns, but may be seen in children ages 6-24 months, because their bones are growing so rapidly.[73] A major source of vitamin D is sunlight and its ultra violet rays, acting on our skins. Vitamin D promotes the absorption of calcium and phosphorus from the gastrointestinal tract to bones. A deficiency of vitamin D makes it difficult to maintain proper levels of those elements in bones,[74] causing them to become weak and soft.

In 2005, scientists from Pennsylvania State University identified a mutation of a particular gene that appears to play an important role in human skin pigmentation. One of the key researchers, Keith Cheng, states that the gene is one of those that appear to modulate "the number, size and density of melanosomes." These are the pigment granules that provide tissues with color that protects against ultraviolet radiation (UV). Too much radiation can damage DNA, which, in turn, would result in numerous harmful effects, including skin cancer. The melanosomes provide this defense by the synthesis and transport of melanin pigments. The gene in question, as previously mentioned, is known as SLC24A5.[75] The mutation is virtually non-existent in Asian and African populations, but found in about 99.9% of Europeans.

It should be noted that geneticists and biologist among others, have ultimately discounted an earlier hypotheses that the dark skin of Africans evolved as protection against sun cancers. They did so despite the fact that light skinned people are many times more likely than dark skinned people to die of skin cancer given equal exposure to sunlight. Their reason for the discard of that opinion was the same as that articulated here earlier for a different reason. Skin cancers, strike later in life, usually past the child bearing, or child siring age, and would not affect continuation of the population. It is for the same reason that many of the maladies associated with lack of Vitamin D can be discounted as having any evolutionary consequences. The reason for the dark skin to begin with apparently remains undetermined.

Cheng and his colleagues discovered that people with the normal form of the gene had brown skin, and that fair skinned people of European descent had the modified form of the gene, an allele, caused here by a mutation, that resulted in fewer and smaller melanosomes. The myriad of human skin tones, it should be noted, is a continuous trait, not limited to a few well defined colors. Though this gene apparently causes skin-color differences, the effect does not extend to melanin deposition in other parts of the body such as the hair and eyes, whose tints are under the control of other genes.

The Penn State team that discovered this allele found that it is responsible for only about one-third of the pigment loss that made black skin white, or pale and essentially colorless. A study conducted in 1997 involved a review and analysis of data from more than 100 populations comparing skin reflection, which is the determination of skin color, to geographic latitude.[76] They found that on average skin reflectance is lowest at the equator, then increases, about 8% per 10° of latitude in the northern hemisphere. By way of example and orientation, cities such as Bordeaux, Milan, Belgrade and Bucharest in southern Europe are all at the latitude that may have been the earliest entry into Europe. They are all on or within a few degrees of the 45th parallel. According to the Penn State formula the skin reflexivity of the immigrants in time would have been 36% greater than in the equatorial area, their point of origin. It all sounds very straightforward, but we might wonder whether natural phenomena are so predictable or, in biological matters, so reducible to formula.

On closer examination it seems to possibly have the earmarks of a "just so" explanation, drafted to explain the otherwise unexplained phenomena of skin tone variations. It does, in fact, run contrary to other studies to the effect that latitude does not consistently predict the average serum 25(OH)D level of a population. The term refers to the accepted test for the amount of vitamin D in the body. According to the authors of the study,

> The assumption that vitamin D levels in the population follow a latitude gradient is especially questionable in view of surveys which have shown

> that UVB penetrating to the earth's surface over 24 hours during the summer months in northern Canada equals or exceeds UVB penetration at the equator. Accordingly, there is sufficient opportunity during the spring, summer, and fall months at high latitude for humans to form and store vitamin D-3.[77]

Though it is doubtful that anyone has a final answer, we continue with our survey of the alleged genetic underpinnings of the color variations.

For one thing, many other genes are involved in skin color or tone, in various ways. One of those genes in particular has also been shown to be a major factor in the lack of skin color of modern Europeans. It is almost always found in European populations but is extremely rare elsewhere. Another has an allele, that is, a form of the gene, found solely in 40 to 50% of Europeans[78] and also linked to light-colored skin in studies of mixed-race populations.

Another[79] has been associated with variations of skin color in African-Americans of mixed West African and European descent. It is estimated to account for 15-20% of the melanin difference between African and non-African populations. This form, or allele, occurs in over 80% of Europeans and Asians compared with less than 10% in Africans. Another inhibits any dark pigment produced by melanin. A mutation of it has been found in roughly 80% of Europeans, 75% of Asians and 20-25% of Africans. Two other mutations have also been linked with skin color variation within European populations and have a similar frequency distribution.[80]

Still another has been shown to account for about 8% of the skin tone difference between African and East Asian populations.[81] It is found in 85% of East Asian samples and is non-existent in European and African samples.[82] Three other genes have been found to be linked to human skin pigmentations that have forms, or alleles, with significant frequencies in Asian populations. While not linked to measurements of skin tone variation directly,[83] they have been indicated as potential contributors to the evolution of skin tone in East Asian populations, which are lighter than that of Africans.

Scientists, despite their admirable expertise, like all humans sometimes fall prey to the comfortable conviction that they have all the facts, thus making their judgments irrefutable. It is hardly controverted however that all of their conclusions in this area are based on statistical studies, complex enough due to many known factors, namely the many genes that must be taken into account, in addition to the factor of diet. There is also the clear possibility of further mysteries of human physiology still not discovered or even suspected.

The team of geneticists at Penn State that discovered the SIC24A5 mutation addressed the subject of the time of its first appearance. Although precise dating is impossible, several scientists speculated that on the basis of its spread and variation the mutation arose between 20,000 and 50,000 years ago. That would be consistent with hypotheses relating it to the wave of ancestral humans who migrated eastward and northward out of Africa beginning about 50,000 years ago.[84]

There is one school of thought, however, that places the age of this mutation at between 5 and 12,000 years ago, which would be within the time frame of the rise of agriculture in the Fertile Crescent.[85] They believe also that the Fertile Crescent was the location of that first change and resulted from the beginning of a diet based on grains, poor in vitamin D, in lieu of the fish and meat diet, rich in the vitamin.

Scientists addressing the problem today state that the deficiency can cause rickets, but often include the further explanation that the danger exists only when dietary supplements, including foods rich in the vitamin, are not otherwise provided.

They mention particularly the fact that Arctic populations such as the Inuit and the Lapp, whose skin is relatively dark despite the high latitude of their homeland, have had a diet that is historically rich in vitamin D. The diet is of course their heavy reliance on fish. As we will shortly see, those groups are not the only examples of the failure of the accepted theory to explain skin tone variations of the world's populations. These other populations, though always relying in part on animal flesh, also have a substantial reliance on grains, but still have skin of various colors, and are not pale.

Foods rich in the vitamin have been abundant over wide areas of our interest for many thousands of years, and all evidence indicates that the cereals and grains supplemented, but did not supplant, animal flesh. We evolved as meat eaters and we still are.

Scientists have also cited the widespread occurrence of rickets in 19th-century industrial Europe. They acknowledge however that whether dark-skinned humans migrating to polar latitudes tens or hundreds of thousands of years ago experienced similar problems is open to question.

Vitamin D, it thus appears, is absorbed from food in addition to that produced by the skin when exposed to sunlight. Hence a deficiency may occur in people who must stay indoors or work indoors during the daylight hours, or live in climates with little exposure to sunlight, remembering however that UVA rays penetrate clouds. The deficiency also occurs in those who are lactose intolerant, or who do not drink milk products. It can also result from following a strict vegetarian diet, as most of the natural sources of the healthy form of that vitamin are animal-based, including meat, fish and fish oils, egg yolks, cheese, and beef liver[86].

In summary, the "natural selection" hypothesis is that, because melanin acts like a sun-block, dark-skinned individuals in particular, may require extra vitamin D at higher latitudes, and that lighter skin color evolved to optimize its production by sunlight in "extreme northern and southern latitudes."[87] Let us see how that holds up.

CHAPTER VIII

EVOLUTION IN OVERDRIVE

Despite the genetics and the statistics the hypothesis must be put to the test, namely how well the theory fits observed facts. We will look first however at the studies and conclusions of other researchers.

Dr. Michael Holick is a Professor of Medicine, Physiology and Biophysics. He is Director of the General Clinical Research Center; the Vitamin D, Skin and Bone Research Laboratory; The Biologic Effects of Light Research Center; and Boston University Medical Center. What he has to say is not in the context of skin color or evolution, but has a bearing on those subjects nonetheless.

According to him,[88] in the summer, UV-B rays from the sun can create all of the vitamin D that we need if we get some exposure on our skin. In the winter, the further away we get from the equator, the less chance we have of doing that. For example, in Boston, one can make all the vitamin D needed in the spring, summer, and fall months, but from about November to February, above 35° north latitude and below 35° south, one can't make any vitamin D at all from sunlight exposure. This is the first, and as far as I can find, the only reasonably reliable estimate of the degree of latitude where such a boundary can be drawn. Hence I feel it could be properly used in examining other areas on Earth for determining adequacy of sunlight for vitamin D purposes.

The diminution of sunlight, and UV rays, is a function of the angle at which the rays strike the particular area of the Earth. The angle of the sun at greater distance from the equator, results in longer passages through the atmosphere than occurs from directly overhead where the sun appears every day at the equator at mid-day. Boston is at about 43°. Presumably the same rales apply to the southern hemisphere except that the corresponding months of less sunlight are May until August. Holick also states however that during these times, the body has to rely on its stores of vitamin D from previous months and/or vitamin D from foods or other supplements.

How much sun is needed? Dr. Holick states that for most people and locations, during the summer, a good amount of sunlight exposure is 5 to 15 minutes on the arms and legs, two to three times a week.

He says that exposure of a person in a bathing suit to a minimal dose of sunlight is sufficient to cause redness, or even irritation of the skin. This amount typically would require no more than 15-20 minutes on Cape Cod in June or July at noon time. That would be the equivalent to taking 20,000 IU (international units), which are 25 micrograms (millionths of a gram) of vitamin D orally. In the absence of any sun exposure, 1,000 IU of vitamin D-3 (the form of the vitamin created by sunlight, either on the human skin or on other animals and eaten by humans) in a day is necessary to maintain healthy levels of 25-hydroxyvitamin D, which, as previously mentioned, is a measure of the storage form of that vitamin present in the blood serum. Much data has demonstrated that neither children nor adults are receiving an adequate amount of vitamin D from their diets or from supplements. Vitamin D deficiency, he continues, is extremely common among all races even in the summer.

A word about vitamin D-3. There is a vitamin D-1, which we can forget about. It does not enter our picture at all. Vitamin D-2 is a derivative of a fungus membrane, which is produced by some organisms of phytoplankton, which are microscopic organisms living in both salty and fresh watery environments. Some are bacteria; some are unicellular organisms whose cells are complex structures enclosed within membranes, known as eukaryotes. Except for mushrooms, land vegetation, as well as any invertebrates, do not produce Vitamin D-2,[89] and the mushrooms require irradiation to produce it, presumably exposure to sunlight being sufficient. Otherwise, D-2 it is available to humans through fortified foods such as juices, milk or cereals.[90]

Vitamin D-2 and D-3 are both absorbed by the intestines and converted into active forms in the body by the liver and kidneys. The active Vitamin helps to maintain calcium and phosphorus balance in the body.[91] Research done by Dr. Michael Holick and others at the Endocrine Section of the Department of Medicine, Boston University School of Medicine, established that Vitamin D-2 Is as effective as Vitamin D-3 in maintaining circulating concentrations of 25-Hydroxyvitamm D.[92] Normally D-2 has no advantage over D-3 except as an antifungal medication to treat diseases like West African sleeping sickness.[93]

One last factoid about vitamin D. It is not really a vitamin, but, being manufactured by the body is a hormone. But scientists would obviously rather call it a vitamin, so also shall I. Now back to Dr. Holick and other physicians.

Dr. Holick also feels that every adult and child needs to take a minimum of 1000 IU of vitamin D along with a multivitamin that contains 400 IU of it. This is substantially more than that presently recommended by the FDA. Its current recommended daily amount (RDA) for vitamin D is 200 IU for all people 50 years old or younger. It increases to 400 IU for people older than 50 and to 600 IU for people over 70.

These quantities however may be outdated. Dr. Robert Heaney, a prominent researcher, served as a member of the Food and Nutrition Board over a decade ago to help establish the RDA for vitamin D. According to him, the RDA of 200 IU was based on how much of it is needed to prevent rickets. This is troubling, he states, because vitamin D deficiency can still cause significant problems at levels that aren't extreme enough to cause rickets.[94] It does indeed, but only rickets would be our concern, concerned as we are now with evolution, and for that target, the smaller amount is apparently adequate for children. The other significant problems mentioned usually come too late in life to be a factor in evolution.

Dr. Heaney's concern was undoubtedly referring to a recent statement in the online version of the American Journal of Public Health to the effect that the high prevalence of vitamin D deficiency combined with the discovery of its increased risks of certain types of cancer suggest that the deficiency may account for several thousand premature deaths from colon, breast, ovarian and other cancers annually. Other maladies have also been associated in part with deficiency of that vitamin, such as heart disease, autoimmune diseases, diabetes, and depression, most all of which conditions usually occur late in life. Its victims are usually past child bearing or child siring age, and probably well past life expectancy of the early immigrants into Europe.

Dr. Ray Sahelian, a vitamin researcher with Duke University acknowledges the dangers of inadequate vitamin D, but warned that:

> High dosages of vitamin D may cause short term or long term side effects. Excess intake makes the intestines absorb too much calcium. Headache, nausea, vomiting, loss of appetite, dry mouth, abdominal or bone pain, muscle pain, fatigue and dizziness are some of the symptoms of vitamin D toxicity. Itching, impaired kidney function, calcification of organs and blood vessels, osteoporosis, and seizures are other signs that develop at the later stages.[95]

The doctor was here speaking of the vitamin obtained from sources other than sunlight. The body has defense against overproduction of the vitamin caused by sunlight, which automatically kicks in when the dosage reaches a certain level. The body however does not protect as well against excessive UV absorption.

Some indication of the lack of impact that diseases in later life have on evolution may reside in the findings that it is in the elderly, who need it most, that the deficiency is most common, as the ability to synthesize vitamin D declines with age.[96] One reason was stated by the Mississippi State Dept. of Health: "As people age, the skin doesn't absorb and process the sun's ultraviolet light as effectively as it does in younger adults."[97]

To the contrary however, Penn State, Institute of Medicine reported in November 2010 that review of nearly 1,000 published studies on vitamin D and calcium confirmed that though they do play a huge role in skeletal growth and good bone health, they could not find any strong evidence that vitamin D protected the body against cancer, heart disease, autoimmune diseases and diabetes.[98]

In the United Kingdom there is no official recommended dose "but grey skies and short days from October to March" mean 60 per cent of the population has inadequate blood levels of the vitamin by the end of winter. The grey skies, it should be recalled, block the UVB rays that synthesize vitamin D, but do not block the cancer causing effects of UVA. The UK Food Standards Agency maintains that most people should be able to get all the vitamin D they need from their diet and "by getting a little sun." But contrary to the opinion of Dr. Heaney, this source claims that the vitamin can only be stored in the body for 60 days, 30 short of Dr. Heaney's three month requirement. Did evolution react with such heavy hand as to remove all coloring from Europeans to counter a 30 day shortage, especially considering the alternative sources in the diet and a very low mortality rate for rickets?

Of considerable importance to us is the fact that there are causes of rickets other than lack of sunlight and improper diet. Babies born before their due dates have been found to be more likely than others to develop rickets. Also breast milk does not contain enough vitamin D to prevent rickets, and infants exclusively beast fed, are thus at higher than normal at risk.[99]

Genes may also be a factor. Hereditary rickets is a form of the disease that is passed down through families. It shows up in a number of ways, such as when the kidneys are unable to hold onto the phosphates, kidney disorders that involve renal tubular acidosis, in children who have disorders of the liver, or who cannot convert vitamin D to its active form. It is most likely to occur in children during periods of rapid growth, when the body needs high levels of calcium and phosphate.

Disorders that reduce the digestion or absorption of fats will also make it more difficult for vitamin D absorption. Obesity impairs vitamin D utilization and obese people need twice as much of it as those of normal weight.[100]

It is further noted, and emphasized that Vitamin D deficiency does not automatically, or even often, cause rickets. On the contrary, the vitamin deficiencies are much more prevalent than the disease. Deficiencies are found in 32% of doctors and medical school students, 40% of the entire U.S. population; 42% of African American women of childbearing age, and 48% of young girls, 9-11 years old.[101]

Dr. Holick's figures are of even more concern. Severely deficient in vitamin D were 76% of pregnant women; 81% of babies born to women who were vitamin D deficient themselves; 32% of physicians and medical students; up to 60% of all hospital patients, and 80% of nursing home residents; 42% of African American women of childbearing age, and close to 50% of girls between ages 9 and 11.[102]

But despite the high rate of deficiency, mortality from rickets is low in comparison to that of many other conditions and diseases. Further its mortality rate does not always correlate to UV intensity. Analysis by Dr. L. Paulozzi of white children in southern U.S. states, where sunlight is greatest, has shown that the rickets mortality ratio was 3.11 deaths per 100,000 children. This compares with 1.91 in the northeast, 1.75 in the Midwest, and 1.04 in the western states of the USA.[103] And if there is any evidence that the survivors of rickets are disadvantaged in procreation, it is difficult to find it in the scientific literature. There is, in any event, according to these figures, no direct relationship between mortality rate from the vitamin deficiency and availability of sunlight.

There is a strong correlation of rickets with osteoporosis. These same ratios were found to be associated with standardized hip fracture rates for elderly white patients in 1986-1993. Because the southern region with its greater amount of sunlight had the highest rates of both rickets mortality and hip fracture, it did not appear that lack of sunlight was a significant factor. However, as both rickets and osteoporotic hip fracture are associated with low dietary calcium intake, Dr. Paulozzi suggests that calcium deficiency may explain both conditions in the southern states, and recommended calcium supplements for young people. Hence vitamin D deficiency and rickets are not synonymous and the presence of one does not always mean the presence of the other.

Hence it hardly significantly affects the survival of any population, which is ultimately the most germane inquiry into causes for evolutionary change.[104]

There is other evidence that lower Vitamin D levels do not necessarily correlate with bone loss. African Americans often have a very low circulating 25(OH)D level. As previously mentioned that is the shortened term for the measure of the storage form of vitamin D present in the blood serum. It is the most accurate way to measure how much vitamin D is in the body. However, those of African descent, despite the low level of Vitamin D have high parathyroid hormone. The parathyroid controls calcium within the blood in a very tight range, and thereby controls how much calcium is in the bones, and consequently, the strength and density of the bones. Some recent studies, including one in 2010, have found evidence that low vitamin D levels among people of African ancestry may be due to reasons other than lack of sunlight or diet. Specifically it dealt with the fact that black women have an increase in the serum parathyroid hormone in comparison to Caucasians.[105]

Significantly, and in corroboration of that finding, those of African descent, though associated with vitamin D levels lower than other ethnic groups have the greatest bone density[106] and lowest risk of fragility fractures compared to other populations.[107] It has also been found that having darker skin and reduced exposure to sunshine did not produce rickets unless the diet deviated from a Western dietary pattern characterized by high intakes of meat, fish and eggs, and low intakes of high-extraction cereals.[108] The Africans of the prehistoric epochs, including the early immigrants to Europe, by all available evidence did have diets of substantially that pattern.

The study also found that the disease continues to be problematic among infants in many communities, especially among infants who are exclusively breast-fed, infants and children of dark-skinned immigrants living in temperate climates, infants and their mothers in the Middle East, and infants and children in many developing countries in the tropics and subtropics, such as Nigeria, Ethiopia, Yemen, and Bangladesh. At first blush, this appears to corroborate the idea of correlation between skin color and rickets.

The authors further state however that the major cause of rickets among young infants in most countries is the fact that breast milk is low in vitamin D and its metabolites, social and religious customs, and the fact that climatic conditions often prevent adequate ultraviolet light exposure. However, they continue, in sunny countries such as Nigeria, South Africa, and Bangladesh, as should be obvious at once, such factors do not apply. Studies indicated that the disease occurs among older toddlers and children and probably is attributable to low dietary calcium intakes, which are characteristic of cereal-based diets with limited healing of the bone disease. Despite the relationship between calcium and the vitamin, there frequently are substantial differences between deficiency in calcium, which is not created by sunlight, and deficiency in vitamin D which is. It has been found, for instance, that nutritional rickets exists in countries with intense year-round sunlight such as Nigeria and can occur without vitamin D deficiency.[109]

Other studies among Asian children and African American toddlers suggested that low dietary calcium intakes in those populations can result in the development of vitamin D deficiency and rickets.[110] In short, the deficiency in calcium can result, in addition to inadequate sunlight or by dietary deficiency, any of the other circumstances listed above.

Though one study shows that Vitamin D levels are, in fact, approximately 30% lower in northern Europe than in central and southern Europe, the difference may be genetically based.[111] In a meta-analysis of cross-sectional studies on serum through the "25-hydroxy vitamin D" (25(OH)D) test, concentration levels significantly, showed no trend in the serum level which correlate with latitude.[112] Also Altitude is seldom taken into account by most authors, but the higher the altitude above sea level, the thinner the atmosphere and its filtering effects and the greater the exposure to UV-B rays.[113]

A number of researchers suggest that the northern latitudes permitted enough synthesis of vitamin D from sunlight combined with food sources from hunting or fishing to keep prehistoric populations healthy. Perhaps, as shown below, modern scientific studies corroborate the high probability for substantial intake of the vitamin in foods available in prehistoric times.

According to the Dietary Reference Intake[114] and the FDA, relatively small amounts of oily fish such as salmon and herring can supply much of the recommended amounts as they thrive on vitamin D rich plankton. Catfish also feed on the minuscule sea life that creates the vitamin from sunlight.

Sardines and tuna likewise provide substantial amounts. Cod liver oil is very rich in sunlight Vitamin D, which is a way to get quickly a healthy dose without being in the sun. One tablespoon of the oil, for those who can handle the foul taste, provides 1,360 IUs. The yolk of eggs also contains small amounts.

Though meat contains generally much less Vitamin D than fish, it does contain enough to be significant in combination with sunshine and other foods. Meat was apparently a substantial part of the diet, not only in the Paleolithic, but also in the Neolithic, the New Stone Age, with emphasis on the lands of the Fertile Crescent.

Much of the evidence for this are the paintings and engravings of animals that were most prevalent, according to the fossil record, in the areas at the time in question.[115] There is, for one instance, evidence both in the fossils and in the paintings in the Gravettian caves of the Paleolithic of the hunting of mammoth, horse, cave bear, bison and other animals. Fishing has been known to humans for hundreds of thousands of years.[116] Indisputable evidence of fishing about 300,000 BP has been found at a site called *Terra Amata* on the Mediterranean coast of France. And during the Late Stone Age in Africa, humans not only hunted, but fished and fowled routinely.[117]

There is one other thing we should know about Terra Amata. It has been established that our ancestors there not only ate fish, but also vegetation. We speak off-handedly of "hunters and gatherers," but, for some reason, picture only the ostensibly romantic calling of the hunters. Gathering meant gathering vegetation, plants, vegetables or fruits. None have any of the healthy form of vitamin D except for mushrooms. Cultivation of grains and cereals has led some to see that as causative of lightening of the skin. It was not the grains or cereals that they blame, but the fact that those foods took the place of animal flesh.

But they certainly did not, no more than did the vegetation that was eaten at Terra Amata. Those folks ate both vegetation and fish, just as all other cultures, whether engaged in agriculture or not, ate vegetation and animal flesh of some kind.

They supplemented their meat diets with some type of vegetation. But none of them developed pale skin except the much later Europeans of the UP, and the Neolithic. But specific lipid biomarkers left inside ceramic vessels that date from the Fertile Crescent in Neolithic times prove that even then, and even there, meat was still eaten.[118]

As the ice began to melt in northern Europe at the end of the ice age, between 18,000 and 12,000 BP. the climate was warming and moistening in the south, but most of the mammals lingered, and remained the humans' chief source of food. Around the peripheries of the ice, were horse, bison and mammoth. In the northern and northwestern part of the continent the reindeer remained the most important source of food.

What about the Fertile Crescent and the claim that the new cereal based diet caused the change in skin color? Even apart from the evidence in the ceramic vessels, it seems to be just as frail as the claim of change in earlier Europe. Between the Upper Paleolithic and the Neolithic was a relatively short period of transition known as the Mesolithic. Frequent subjects of their art, and apparently their sources of food, were elk, reindeer, bears and sea mammals.[119] The slightly later Neolithic sites of the lower Danube were mostly villages and towns where the settlers raised grains and kept herds of cattle sheep, goats and pigs.[120]

A bit later came a culture known as the Lake-dwellers. They built habitations of logs supported on piles placed in shallow waters around the shores of lakes extending into the Italian Alps, Austria, Bavaria and eastern France around the periphery of the Alps. This Alpine Neolithic, though agricultural, raised cattle, goats, sheep, pigs, and dogs, and they hunted extensively for deer which abounded in the nearby mountains.[121]

Hence the picture so off-handedly painted of the Neolithic people, suddenly abandoning the hunting of animals for meat, is not supportable. Neolithic populations continued not only to hunt, but also domesticated cattle for meat and daily products.

How much vitamin D could they have obtained in addition to any other health benefits? Comprehensive figures have been compiled by the Marshall Protocol Knowledge Base, a Canadian organization of the Autoimmunity Research Foundation.[122] What the adults ate, particularly the women, is important for the fact that their own sufficiency of vitamin D had consequences for their offspring.

The Marshall Protocol gives highly detailed data on the vitamin values of a seemingly inexhaustible list of meats and ways of cooking. Most of the figures involve either 75 or 90 grams of meat, tiny portions hardly enough for a child, but containing typically about 18 IUs of vitamin D. These values are not as high as those for most fish and many other foods. But a diet including meats, together with whatever sunshine is available can certainly afford enough IUs to protect most of a population, at least for long enough to spawn another generation.

And not all physicians agree with the assertion that there is an optimal level of sun exposure.[123] Some argue that it is better to minimize sun exposure at all times and to obtain vitamin D from other sources.[124]

And finally, we turn again to the meeting of the American Association of Physical Anthropologists in March 2007. The anthropologists argued the validity of the more recent dates, 5 to 12 thousand BP for the beginning of pale skin, and brushed aside the conventional wisdom supporting the early date of about 40,000 BP. The conclusion was based on a report that evolution of a gene for skin color indicated that the pale skin was acquired perhaps only 6 to 12,000 years ago.

An article in sciencemag.org, in April of that year by Ann Gibbons explained that the range of dates was the result of imprecision of the methodology for age determination.[125] She also states that these findings contradict a long-standing hypothesis that modern humans in Europe "grew paler about 40,000 years ago, as soon as they migrated into northern latitudes."

The researchers, in fact, suggested that the northern latitudes permitted enough synthesis of vitamin D from sunlight, combined with food sources from hunting to keep populations healthy. Only when agriculture was adopted was there a need for lighter skin to maximize the synthesis of vitamin D.

The aplomb, the ease, with which the anthropologists tossed aside the prevailing dicta about lack of sunshine as causing inadequate Vitamin D in the immigrants into northern Europe, seems highly significant, no matter what the ultimate fate of their main thesis will be.

CHAPTER IX

DOES EVOLUTION PLAY FAVORITES?

Among the areas we will examine to determine the validity of the standard theory are those of the Eskimos and the limits in northern Canada and Greenland, the Lapps in northern Scandinavia, and the Selknam in Tierra del Fuego in the very southern area of South America. All are well outside of the tropics or subtropics, the zones where sufficient UV from the sun is always available due to its daylight position either directly overhead, or at not too step an angle. In these lands, relatively distant from the equator that we will examine, it is at a sharper angle causing more UVs to be filtered out by the atmosphere, but the inhabitants are still dark skinned.

We turn first to the Eskimos of Alaska. A good start would be Illustrations 12, 13 and 14 below. All are native Alaskan Eskimos, part of population living for many thousands of years in that land, now a state of the U.S.

PALE SKIN, GIANTS, AND THE GREAT TRANSITION

Illustration12: By Katie Spielberger CCW Staff Writer; subject of portrait (September 2008): Morgan Fawcett, Juneau, a Native Alaskan musician

Illustration 13: Four native Alaskan teenage girls (2007). Credit: Chelsea A. Kerrington

Illustration 14: Alaskan man
Credit: http://en.wikipedia.org/wiki/Alaska_Natives

The southernmost part of Alaska, except for the Aleutian Islands, is at 60° north. Just about 6° further north is the Arctic Circle that cuts across the state at about its middle. Hence during long stretches during the winter months the inhabitants see no sun at all. Below the circle the inhabitants do well to see a few hours of sunlight at midday, and then at a sharp angle, filtering much of the UV rays.

Based on archaeological evidence people have lived in Alaska for at least 15,000 years. The ancestors of Alaska's Indians, Eskimos, and Aleuts, probably entered North America from Asia across a land bridge known as Beringia which was about 1,000 miles wide, connecting Alaska and Siberia. Some might have come earlier over a smaller land bridge, possibly as early 25,000 years ago, before the glacial maximum, but when sea level was already much lower than it is today. That early date is claimed for bone objects with markings which some archaeologists say were cuts made by humans, but by others to have been by animals. They were found at blue Fish Caves in the Yukon.[126] Hundreds of years passed as people lived on the land bridge and moved across it from Asia to North America. Alaska was the first part of North America these people came to and many settled there permanently.

Most archaeologists think that there were three major migrations of people to Alaska, involving different groups of people and occurring thousands of years apart. The first occurred at least 15,000, years ago. These early immigrants are the ancestors of most Indian tribes in North and South America.

The second group of immigrants consists of the ancestors of the Tlingits, Eyaks, and Athabaskans of Alaska and the Navaho and Apache people of the American Southwest. They moved from the northeastern forests of Siberia, at equally high latitudes, perhaps 14,000 to 9,000 years ago. The last group of immigrants moved into Alaska about 10,000 to 6,000, and some perhaps as recently as 4,000 years ago. They came from the coast of northeast Siberia and are the ancestors of the Eskimos and Aleuts.

The oldest sites of uncontroverted human occupation in Alaska date back 11,000 years. One early site is Onion Portage, located on the Kobuk River in Northwest Alaska, at 67° north. Different groups of people lived at this site for thousands of years. The tools excavated from the most ancient soil layers are similar to those found in Siberia. They are the kind of tools that would have been used to cut meat, scrape hides, work wood, or make weapons.

It appears that evolution has had plenty of time in which to work its magic and healing powers. These immigrants to the new World were getting a lot less sunlight, and UV rays, than were the immigrants to Europe. Why did evolution bestow supposedly protective pale skin on Caucasians and treat the Eskimos and other native populations of Alaskans with such a half-way job? It starts to sound more like the unpredictable ways of people rather than the laws of nature.

We turn to the Lapps. They are residents of the very northern portions of Norway, Sweden and Finland, and the entirety of the Murmansk Oblast. That oblast (a political subdivision of Russia) is the northernmost region of Northwestern Russia. It is the Kola Peninsula, which borders Finland's Lappland to the west and Norway's "Finnmark" Region to the northwest. The Arctic Circle on the map below would pass the southern tip of the oblast.

Illustration 15: Map showing all of Lappland. All parts of Lappland (sometimes spelled as Lapland) are either right on, just barely below, or well above the Arctic Circle, the places on Earth that receive less UV rays than any other except the Antarctic.

Winter usually begins in mid-October, which is at least a month earlier than other parts of Finland, and is the longest season, lasting up to 200 days. The freezing temperatures can fall to as low as 2-5°F. After the onset of winter, the snow falls on open ground for around two weeks and the snow cover gets thicker right through to March. Ultimately the issue is UV rays and sunlight. However apart from the lack of sunlight, and the sharp angle when it is present, snow on the ground comes from clouds, and one type of UV rays, of course, do not penetrate clouds no matter the time of day or season of the year. Despite these circumstances, farming appeared in north Scandinavia as early as about 6,000 BP.

Let us see how well evolution has taken care of the Lapps *a la* the more southern Europeans.

The original settlers of the area were a population now known as Sami. In prehistoric times, the Sami people of Arctic Europe lived and worked in the northern arctic region of Finnish-Scandinavia and Russia for at least 5000 years. Petroglyphs and archeological findings such as settlements dating even earlier, from about 10,000 B.C., can be found in the traditional lands of the Sami, left by a population known as Komsa. Archaeologists are convinced that the Sami are a continuation of the same population as the Komsa. It is hypothesized that the Komsa followed receding glaciers inland from the Arctic coast at the end of the last ice age, between 15,000 and 12,000 years BP, to the modern Finnmark area of Norway to the coast of the Kola Peninsula, newly opened up for settlement. The Sami are the earliest ethnic group in the area, and are considered an indigenous population.

Once again there has been plenty of time for their skin to have turned pale, but for reasons unexplained by the evolutionary biologists who attribute light skin solely to lack of UV rays, it has not.

Illustration 16: Typical Lappland Boy: Lappwoman from the north of Lappland: bottom Lapp-woman; Lapp from Norrbotton, Sweden.

Illustration 17: Photo:
Credit: *1925:* Richard Fleischhut

Illustration 18: *A Sami (Lapp) family in Norway around 1900.*
Photo: Picture credit: Library of Congress;
http://www.environmentalgraffiti.com/featured/
before-fall-reindeer-people/18225

PALE SKIN, GIANTS, AND THE GREAT TRANSITION

Illustration 19: Warm Hearts of the North, Sami mother, *pre-1923*
Photo: Borg Mesch

Illustration 20: wedding, @1930

Illustration 20 will require close examination, but the skin of the bridal party and attendees is certainly dark, not pale.

The Inuits are inhabitants of northern Canada and Greenland as well as parts of northern Siberia. They are the descendants of what anthropologists call the Thule culture, who emerged from western Alaska, after crossing the land bridge from Asia, around AD 1000 and spread eastwards across the Arctic. They displaced the related Dorset culture, the last major Paleo-Eskimo culture (in Inuktitut, called the *Tuniit*).

Inuit legends, as do many others, speak of the Tuniit as "giants", people who were taller and stronger than the Inuit. By 1300, the Inuit had settled in west Greenland, and they moved into east Greenland over the following century.

Faced with population pressures from the Thule and other surrounding groups, such as the Algonquian and Siouan to the south, the Tuniit gradually receded. They were thought to have become completely extinct as a people by about 1400 or 1500 AD. But it had been sometime in the 13th century that the Thule culture began arriving in Greenland from what is now Canada.

The Inuit of Canada and Greenland circulated almost exclusively north of the "Arctic tree line", the effective southern border of Inuit society. The most southern "officially recognized" Inuit community in the world is *Rigolet*. The descendants of the southern Labrador Inuit continued their traditional semi-nomadic way of life until the mid-1900s. Both Finnish and Sami minorities continue to this day to maintain their cultures and identities.

In West Hudson Bay, the location of Illustration 20 below, the winter blizzards abound and temperatures potentially reach as low as—53° Fahrenheit.

PALE SKIN, GIANTS, AND THE GREAT TRANSITION

Illustration 21: A Group of Inuit, West Hudson Bay, Canada
Credit: Captain George Comer: http://mysticseaport.wordpress.com/tag/inuit/

Illustration 22: Nine Inuit pose for a photographer in 1913,
***Credits: Library of Congress Prints and Photographs;
Photograph by Northern Ventures Ltd., New York, 1913***
http://www.windows2universe.org/earth/polar/inuit_image_gallery.html

Illustration 23, 1897:
class picture from the limit school in South Head, Siberia.
Library of Congress Prints and Photographs,
Photograph by F.D. Fujiwara
http://www.windows2universe.org/earth/polar/inuit_image_gallery.html

Let us look more closely now at one particular population of American Indians who, like the Eskimos, Lapps and Inuit, were not favored with pale skin, despite their cloud covered home far from the equator. Like the others, it was a group that may exemplify most of the reasons advanced to contradict the standard theory. They are the inhabitants of the very south of the South American continent, an area known as Patagonia, in particular, on the island known as Tierra del Fuego. The Island is cut off from the mainland by the Straits of Magellan. Not the least important fact about the island from our perspective is its location at about 57° below the equator.

The inhabitants were known as the Selknam, or the Ona, and were believed to be one of the most primitive populations on Earth, having occupied that homeland for at least 12,000 years. The inhabitants are extinct now, but until the early 20th Century were occupants of one of the entire southwest part of that island. It is quite cold there, and very windy, winds of about 40 mph being not unusual, with gusts up to about 60. But neither the cold nor the wind is as germane to our interest as the perennial cloudiness, violent showers, storms and powerful hurricanes.

It is interesting in this connection to recall that clouds, though blocking UVB, do not bloc UVA rays, thus permitting some synthesis of Vitamin D even in overcast or rain.

The largest animal on the island is the guanaco, a wild relative of the lama and camel. Water birds also abound as do bountiful fish in the cold waters, and sea mammals including whales and various species of seals. Of the three populations on the island. The Selknam did not fish; the other two did, even though not with fishing hooks or nets So from where do the Selkmans get their vitamin D? Some of it came from the guanacos they hunted. But despite the lack of fish and the distance from the equator, their skin remained dark.

We have two group photos (Ill 24 and 25), probably taken in the late 19th or early 20th Century from which we would not draw any far reaching or broad based conclusions, but they are old pictures and not selected for their healthy and mature appearances. But nature (aka evolution) has obviously neglected to furnish the Selknam with the same benign attention showered upon the Europeans who generally had more sunlight and ability to utilize more resources than did their cousins on the other side of the world. Why did the natural processes of evolution not work for them as it did for the Europeans?

Granted their skin is not as dark as that of the Africans, it has, for reasons apparent from the photographs, been termed "swarthy:" According to one writer, "Swamps, matted roots, grass tufts, and dark earth feature in the creation stories reflecting the topographies of the Selknam domain and genitalia are formed from the very mud, thus explaining the Selknam's swarthy skin color."[127]

At the time of the first European settlement, in the 1880s, they numbered about 3500, and were said to be fearless and expert hunters who used bows and arrows with great skill.

Illustration 24: Selkman Family Group

Illustration 25: Selkman women

By the Late 1920s, they had been reduced to about a hundred, and those survivors soon succsuccumbed to acculturation.' The last two of the Selknam died in the late 1960s. But the cause of obliteration of this population had nothing to do with lack of vitamin D in general, or of sunlight in particular. Nor was it caused by their diet. It was a case of gof genocide, for which neither they nor the 'civilized' Europeans undoubtedly had a word in their vocabularies.

It started mainly with controversy over the island's sheep, introduced by the European settlers who arrived in 1880. They fenced off large areas and forbid entrentries by the natives. The Selknam had little concept of private property and less for cfor confiscation of their land. The controversy escalated with the discovery of goldgold and ended with Caucasian headhunters who were paid a bounty for the heahheads of Selknam they had killed, with bonuses for pregnant women. The heads werwere sometimes shipped to museums and individual collectors. Poison meat was also given to the Selknam in furtherance of the plan to rid the settlers of the pesky natives.[128]

For the growing evidence that neither the late nor early dates of change of skin tone can be entirely placed on lack of sunlight, we need to look further at an interesting study. It is argument for the hypothesis of the recent change, something later than about 12,000 BP, and it comes from a scholar named Frank W. Sweet. It and its ramifications need a separate chapter.

CHAPTER X

A CASE FOR RECENT CHANGE IN CAUCASIAN SKIN TONE

SOME SUPPORT FOR THE EARLIER development of the pale white skin of Caucasians may have been introduced by a scholar named Frank W. Sweet.[129] It must have been unintentional no doubt as he was arguing for the later development, around 5500 BP. It was, for the most part, a well-reasoned article published in December 2002. Sweet is a researcher who thinks outside the box, or at least to a certain extent. Odds are he is one who probably never has, nor would ever, think in terms of extraterrestrials. Nonetheless his logically presented article furnishes us, in my opinion, with more circumstantial evidence for the extraterrestrial hypothesis. It is in the form of a puzzle to which, for our point of interest, he proposes one, and only one solution. Though admitting it could be disproved by certain evidence, he may not be aware that perhaps it already has been disproved by early art.

Sweet sets the puzzle out clearly and early: "Since the last glacial maximum [20,000-18,000 BP], Europeans developed fairer complexions than any other group on earth, fairer even than others at the same or higher latitudes." The other piece of the puzzle that doesn't fit in with the standard picture of world skin tone concerns, he says, the equatorial Native Americans who have significantly lighter skin than the Africans and Melanesians who also live in the equatorial zone. Sweet's explanation for this second group is interesting, but has no bearing on our point of focus, namely the light skin of Europeans. The reasons he gives for each are separate and bear no relationship to the other. We will be concerned only with the Europeans.

He claims that the most "eye-catching feature" of maps of skin tone is that "the lightest complexion on earth is native only to the region within 600 miles of the Baltic and North seas. The feature is unique on the globe."

Included within the 600 miles from the Baltic and the North Sea would be all of France, down to the Mediterranean, and Germany, and well as all or parts of other European countries and eastern Russia. Further emphasizing its uniqueness, he continues, is the fact that blondes are also native only to the same region. "The two minor exceptions are the fair-skinned but brown-haired people of Bordeaux and the blonde but swarthy descendants of the Volga Rus."

Sweet examines, and ultimately accepts the majority view of the significant points of the migration out of Africa, which he pegs at between 60,000 and 50,000 years ago. His shorthand reference to that migration is OOA2 (out of Africa 2), it being the second major exodus from Africa, the first occurring about 2 million years ago. That was long before modern humans of the UP appeared. He states that the OOA2 hypothesis is to the effect that the modern humans totally replaced older species, Neandertals and Homo erectus and their offspring races included, with little or no hybridization. He addresses the "large minority" opposition to that position, and their claim that there could have been gene flow and interbreeding after the OOA2 dispersal, and concludes that the molecular evidence to the contrary is too convincing.

He then addresses a "notion" that lightening of skin around the Baltic European area took place after the last glacial maximum, not before. He supports it with Magdalenian cave art, The Magdalenian period being the most recent of the UP in Europe, from about 16,500 to about 11,000 BP. Scientific evidence does indeed support the conclusion that, practically all of the paintings by the Magdelenians date from after the last glacial maximum (20,000 to 18,000 BP), though the evidence upon which Sweet relies to show dark skin in those paintings is weak. He relies on art depicting dark brown bowmen shooting medium brown deer.

That reliance is misplaced. The cave murals at Lascaux have been dated to the Solutrean-Magdalenian period 21,000 BP to 10,000 BP with the earliest art dating from 19,000 BP. It is indeed unfortunate that the prehistoric painters paid little attention to the visage of their fellow humans. Accurate color and form portrayals of humans could, 16,000 years later, show us so much, though it should surprise no one that they did not think of that. Despite their highly accurate representations of animals, the humans, with the exception of those at one site, La Marche, are mere stick figures with lines for arms and legs, a line or a box for a torso, a circle for a head, often with no facial features shown, or sometimes with a slit for a mouth, and dots for eyes. That one exception, La Marche, requires a chapter of its own, which will be Chapter XII.

Unlike the beautifully painted animals, the lines and circles representing humans are almost always black, brown or grey. That is true throughout France and Spain. In the cave paintings even the darkly colored deer, contrary to the humans, show much evidence of attention to the animal's skin tone as well as accurate depiction of the body shape.

There is a theory advanced by some that not all of the dark skinned peoples at high latitudes had enough time in their new habitat to develop the light skin of the northern Europeans. In support, Sweet traces the time line of the migrations of various populations. We have already seen that there is evidence of humans in Arctic Russia 40,000 years ago. In this connection it is interesting to note that the skill of the Lapp population, inhabitants of northern Scandinavia, has been described as light brown[130] or brown.[131] We also have seen representative illustrations in Chapter IX.

The same source says that the hair of the Lapps is smooth, straight or wavy, dark brown or black. Facial hair is said to be insignificant. The nose is short and elevated. The narrow nasal bridge is straight or concave. The cheekbones protrude. Orbits, bones around the eyes, are small and low. The eyes are narrow, and colored dark brown or black. The lips are narrow and small. The chin is pointed, and the lower jaw is wide and low.[132] Hence, even though they had time enough to develop some Caucasian characteristics, some Negroid traits remain, and the skin color did not become pale. There was obviously no evolutionary necessity that they do so.

Sweet concludes: "If Europeans had enough time to become pink, then so did Asians, Inuits, Aleuts, and Saami," and he identifies the meat that each group ate. "Molecular evidence gives Europeans no more time on site to develop their world-unique complexion than it gives to light brown Asians at European latitudes" So, he says, the question is: why are Europeans pink? His comment: "Some obviously visible regional adaptations do not seem to have had enough time to become fixed, and yet there they are. Other expected adaptations have had just as much time to unfold, and yet there they are not." Here Sweet does not do his theory justice as most recent biological research supports very rapid evolution in certain circumstances.

Sweet rejects the position of some that Vitamin D synthesis by exposure to sunlight alone suffices to produce changes in the global pattern of skin tone: "This sole cause hypothesis has not withstood scrutiny." He cites many of the same figures of the USDA previously cited by us above. He concludes from the UP populations' cave art and artifacts that their diet was well within the range of sufficiency, especially when Vitamin D from sunlight was added in. It is a valid insight. It is not as though sunlight was switched off in the higher latitudes, except above the Arctic Circle and then for relatively short periods except at or near the poles. This, says Sweet, left the migrants with light brown or beige complexion that is common to everyone above the 55th parallel—except the Europeans.

But the diet of the Europeans changed with farming, about which he says this:

> European agriculture began about ten millennia ago in the Near East and spread to the Baltic by five millennia ago. As in Asia, Africa, and America, the advent of agriculture saw a dietary shift from meat to grains.

The word "shift" is undoubtedly a gross overstatement, but possibly necessary to his thesis. He continues:

> This reduced dietary vitamin D intake among farming peoples and so perhaps lightened their complexions slightly via the paleness adaptation. It was probably not significant outside Europe because domestic grains (corn, wheat, oats, sorghum, millet, rice) do not grow without intensive modern agricultural techniques above about 55° of latitude. Higher latitudes are just too cold—the growing season is too short—to let crops compete successfully with herds as food source. Consequently, even post-Neolithic high-latitude peoples continued to have a diet rich in meat (and so, vitamin D). These include the Inuit (seagoing mammals), Aleuts (fish), Saami (reindeer), Mongols (horses), and Native North Americans (bison).

In short, he believes that lack of sunlight and insufficient dietary vitamin D are both necessary to get the pale skin of Europeans. If it was too cold for these populations, Inuit, Aleuts, Saami, Mongols and too cold for Native Americans to grow grains, why was it not too cold for the Europeans? His answer:

> Only one spot on the globe enables economically competitive grain production above the 55th parallel. It is where the warm Gulf Stream washes into the North and Baltic Seas, keeping temperatures moderate despite dim near-Arctic sunlight. Around the planet, only circum-Baltic farmers could switch to a grain diet devoid of vitamin D, in a place where sunlight also lacked UV. And so, the extreme of the paleness adaptation is found only within 600 miles of this unique spot on earth.

In short, Sweet sees grains as the culprit. It was the cultivation and consumption of grains which, he claims, largely took the place of vitamin D rich meat and fish that turned our skins pale as opposed to merely lightening their complexions. But theories and abstractions aside, we must once again look at the facts on the ground.

One of the facts is that there were other parts of the world, far from the equator, where people were also growing grains, or other foods without vitamin D, thousands of years ago. It is not deemed necessary that they be further than the 35th parallel, north or south, to qualify as "far from the equator." That was the parallel mentioned by Dr. Michael Holick, cited in Chapter VIII, who stated that from about November to February, above 35° north latitude and below 35° south, presumably June through August, one can't make any vitamin D at all from sunlight exposure during winter months. He used as an example the city of Boston at 43° north.

Sweet says that the Baltic is the one exception to his statement that domestic grains do not grow above the 55th parallel without intensive modern agricultural techniques, as the weather is too cold. The reason for this unique exception is that the North and Baltic Seas are warmed by the Gulf Steam. That much is true. But he has not favored us with any source for his inference that either the use of grains or cereals in Europe, or pale skin originated in the Baltic area, and I have been unable to find any. The evidence seems to say otherwise.

We have already seen that these Europeans from central and southern Europe continued to hunt through the Mesolithic and Neolithic. In southern Europe the prevalence of sunlight may make that irrelevant, but not so in central Europe.

There is substantial evidence for farming in central Europe earlier than in the Baltic. A crucial question, which became the subject of intense study by a team of highly reputable scientists, including geneticists, archaeologists, and anthropologists, is the extent to which Europeans are descended from the first European farmers in the "Neolithic Age 7500 years ago or from Paleolithic hunter-gatherers who were present in Europe since 40,000 years ago." They analyzed ancient mitochondrial DNA (mtDNA), which is inherited from mothers only, from 24 Neolithic skeletons of European farmers. The skeletons came from various locations in Germany, Austria, and Hungary. They found that these first Neolithic farmers did not have a strong genetic influence on modern European female lineages, and that their finding lends weight to a proposed Paleolithic ancestry for modern Europeans. The results were published in 2005.[133]

The scientists pointed out that agriculture originated in the Fertile Crescent of the Near East about 12,000 years ago, from where it spread via Anatolia, present day Turkey, all over Europe. Archaeological cultures mark the onset of farming in temperate regions of Europe 7500 years ago. These early farming cultures originated in Hungary and Slovakia, then spread rapidly as far as the Paris Basin and the Ukraine. The readers' attention is invited to the top map below. Illustration 26, for orientation as to the location and extent of the Paris Basin. According to the scientists, "The remarkable speed of the LBK expansion [people from Hungary and Slovakia identified by their line decorated pottery] within a period of about 500 years, and the general uniformity of this archaeological unit across a territory of nearly a million square kilometers [bottom figure. Illustration 27], might indicate that either the spread was fueled to a considerable degree by a migration of people or that local European hunter-gatherers had shifted to farming without significant mixing of genes with the migrant farmers."

From our narrow viewpoint, it is immaterial which of the two scenarios is correct. We are interested primarily in the date of the first farmers in central Europe. From the two illustrations below it is apparent that agriculture in central Europe preceded that of the Baltic by about 2000 years. But the findings also make it evident that white skin in central Europe, Sweet to the contrary, could not have come from either the Baltic or the Fertile Crescent. The genes are those of the European occupants from 40,000 BP, the hunter-gatherers. How growth of grains spread through Europe from the Baltic, we get no hint from Sweet

The findings, and the archaeological record, according to the scientists, "confirm that the first farmers in Central Europe had limited success in leaving a genetic mark from the female lineages". This is in contrast to the success of the Neolithic farming culture itself, which quickly spread all over Europe, as the archaeological record demonstrates. "One possible explanation is that the farming culture itself spread without the people originally carrying these ideas. This includes the possibility that small pioneer groups carried farming into new areas of Europe, and that once the technique had taken root, the surrounding hunter-gatherers adopted the new culture and then outnumbered the original farmers." But there is no evidence of influx to central Europe from the Baltic.

Illustration 26: Map showing the location and extent of the Paris Basin

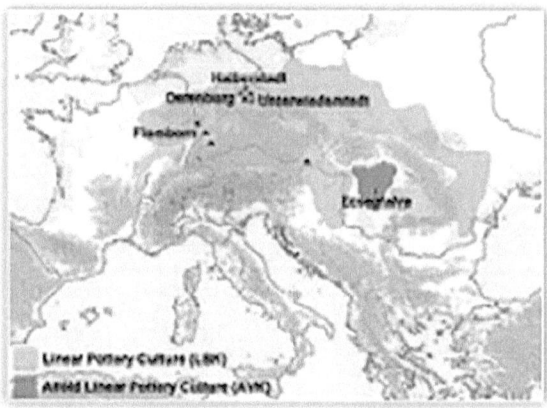

Illustration 27 Geographic range of the first Central European farmers.

The orange and red areas indicate the widest distribution of the earliest Neolithic farming cultures LBK and AVK after 7500 years before the present.

Grains and other foods without vitamin D, however, can be, and have been, grown above the 35th parallel by some very ancient peoples with very large populations, and they do not have the pale skin of Europeans. True they also ate meat and fish, but so did the people of the Baltic and just about everyone else in all times and places. Who are they? We will look at two major areas.

CHAPTER XI

RULE BREAKERS

THERE ARE FOR ONE EXAMPLE the Native North Americans. The Native Americans discovered that wild seed plants growing along river floodplains could be controlled; that plants could be harvested and used as food, with seeds stored and replanted in prepared garden plots the next year. Four indigenous plants underwent this transition to full domesticates, with clear physical changes taking place in their seeds. Three additional cultigens appear as food crops as Native Americans began to harvest these previously wild sources. The highly nutritious seeds from these seven plants could be variously boiled into cereals, ground into flours, or eaten directly.

Each of the seven plants involved, named chenopod, marsh elder, squash, sunflower, erect knotweed, little barley, and maygrass, had its own particular course of development. Most began as wild plants growing along river floodplains that Native North Americans first gathered and used. They gradually brought these plants under their control as they harvested them and planted their seeds the following year.

By 4,000 BP, there is evidence of this indigenous crop domestication occurring over a broad geographical area, on lands today known as the states of Tennessee, Arkansas, Illinois, Kentucky, Ohio, Missouri and Alabama. Of these, all but Arkansas and Alabama can be considered entirely, or substantially all above the 35th parallel. About half of Arkansas would be included, but almost none of Alabama.

After a slow beginning for each crop, the overall shift to domestication occurred rather abruptly, with several spring and fall crops introduced together, some high in oil and some in starch.[134]

There are undoubtedly many other areas of the New World where foods without vitamin D were grown and eaten, many more sites than those for which we have evidence. One site for which we do have evidence is the Four Corners area in Utah, dating to about 3000 BP. It is located at 37° north. The Anasazi successfully developed agriculture there.[135] In what is now Oklahoma, there is evidence that about 5000 BP wild plants were consumed by the native populations. In Kay County, charred Onion bulbs were recovered, and about 3000 BP there is evidence in Oklahoma of plant domestication.[136]

More generally, it was in the arid Southwest that some of the earliest farming societies developed. The predecessors of the Pueblo and Navajo Indians were able to thrive in the desert by developing, among other things, complex irrigation systems for farming. Around 3,000 years ago, however, the climate grew even drier, killing off many of the region's wild game and vegetation. A group of people known as the Mogollon, living in permanent villages along the rivers of eastern Arizona and western New Mexico, responded to this change with increased farming by raising beans, squash, and corn. Almost that entire border is north of the 35th parallel.

Around 2700 BP, other groups of people, known as the Adena began to build large mounds and earthworks in southern Ohio. They lived in small villages and supported themselves by, gathering wild plants, and farming, in addition to hunting and fishing and some trading. The Indians were the first people to cultivate some of the world's most important agricultural crops, including chocolate, corn, long-staple cotton, peanuts, pineapples, potatoes, rubber, quinine, tobacco, and vanilla.[137]

Northeast Asia, including central and parts of northern China is another case in point. Its agricultural origins are usually traced to the loess terraces of China dating back to about seven millennia BP, or between 6,000 and 7,000 BP. A few crops such as foxtail millet and broomcorn millet are thought to have had their origins in northern China. Anything qualifying as central or northern China would be above 35° north. Through subsequent diffusion to the east, these crops joined with other cultigens and ultimately reached Japan about 3,000 BP. Limited gardening seems to have existed in Japan prior to this, which raises the possibility of indigenous Japanese plant domestication and husbandry. Barley and wheat from western Asia and rice from tropical Asia were added to some northern Asian corn complexes at some time before 3,000 BP.

The earliest rice in China appears to date to 5000 BP, in the northern area, about 500 years later. Barley and wheat are found in China by 3000 BP. Oxtail millet is reported from a Yangshao site c. 6000 BP. Banpocun, sometimes known as Shaanxi, was once part of the Yangshao, a culture notable for many things, including mortars and pestles for grinding grain which have all been found and identified there. It is at Latitude 35.138°.[138]

Wheat appears by 2300 BP. There is evidence of wild rice and indications of its possible consumption about 10,000 BP, but not above the 35th parallel.[139]

Foxtail millet was in southwest Japan in the early third millennium BP. Japan is almost entirely above 35° The Sakushu-Kotoni site is at 43°; Kenli at 38°. Broomcorn millet may have been present at the same time; there is evidence that this millet was in Korea in the early Bronze Age (about 4000 BP). Korea is entirely above the 35th parallel, extending to the 45th. Melon seeds are abundant from the late third millennium BP in Japan. Several taxa in the collection in Hokkaido, Japan's northernmost major island, are evidenced even earlier. Vigna, Perilla, and Cannabis are reported from the fifth millennium.

Barnyard grass is an important component of a 4000 BP Middle Jomon assemblage from southwest Hokkaido, the northernmost of the Japanese major islands, and the caryopsis size increase of this population suggests that domestication was occurring. Barnyard grass caryopses is a small dry one-seeded fruit. From the Saknshu-Kotom River site all are wild size.[140]

In 2006 there was published by the Center for East Asian Studies of the University of Pennsylvania, a study entitled "Sino-Platonic Papers."

It began with the unequivocal statement that "Cereal Agriculture was the necessary basis of any civilization in early ancient times." That seems well said. Even in those cases where we have little research or evidence, there are hints that agriculture existed in some degree. It is well-nigh certain that all humans ate meat, and almost as certain that all civilizations grew grains and cereals of some kind, or foraged for them. Whatever the results of their foraging, there was undoubted very little if any vitamin D. No fruits, and only mushrooms among vegetables have any of consequence. The import ant fact is that, in addition to animal flesh, people were eating foods that did not contain vitamin D. The other important fact is that their complexions did not turn pale, and it is difficult to see why lack of the vitamin from grains in Europe would have had such an effect.

The papers just mentioned contain much of value to us, a considerable amount of which was reported in earlier research. According to the authors, from about 9,000 to 4,000 BP many sites along the middle reaches of the Yellow River, and the middle and lower reaches of the Yangtze River gave rise to agricultural civilizations. Each of the two areas consisted mainly of separate kinds of cereals, millet along the Yellow, and rice along the Yangtze. The study was not concerned with latitude or with skin tone, neither of which was of any import to its subject, but which is important to ours.

The Yellow River empties into the sea near Beijing at about 40° north and there are many stretches of the middle part of the river's length of greater latitude, though some are lower. The Yangtze does not run above the 35th parallel for any of its length.

Hence we have in those areas along the Yellow River sites with significant distance, that is, 35° or more from the equator and the existence of agriculture and grains thousands of years ago. But the Chinese populations are not pale, there or elsewhere, a matter sufficiently known and requiring no photographs.

Though we have no evidence of agriculture in the far north of china, we can look north of that country to Mongolia. It is mostly above parallel 45 and extends to the 50th. The origin of agriculture in Mongolia has been found to date back to the Stone Age and the earliest findings related to agriculture are the stone hand mill and mortar found in eastern Mongolia. Settled communities are first evident between 9000 and 8000 BP in Inner

Mongolia and in the Loess Plateau drained by the Huang He (Yellow River) system and other rivers such as the Liao in northeastern China. In all these areas, people were moving toward agriculture by 8000 BP.[141]

Sweet says that the main objection to his hypothesis is its recency, feeling that five or six thousand years is too short a period for such a genetic change. We have seen just how fast the change can be in accordance with epigenetic processes; however it is probably of no importance to the issue here. He mentions two or three factors that could, he claims, shorten the time for the changes to occur. Since they do not, for our purposes, matter at all, we need not discuss them. I feel confident that time has been adequate, and is not an issue here. He acknowledges that the beginning of any such development, according to his theory, could not be earlier than 10,000 B.C., the time that farming was first introduced in the Middle East.

He has creditably knocked out the usual explanations, such as lack of sunlight as the sole cause for the unique paleness of European skin. The only support left standing is the lack of Vitamin D in the diet beginning with the Neolithic and farming. His hypothesis can be disproved, he says, but in only one way he can think of:

"Ultimately, it all depends on evidence. The hypothesis presented here will be contradicted when someone finds evidence as early as Magdalenian cave art, that Paleolithic Europeans were as fair complexioned as Neolithic Europeans."

But, even granting that there was time enough for the change in skin color to spread through Europe, there is no suggestion from Sweet as to why the change would spread through that continent. Yes, evolution, the science of epigenesis considered, can admittedly happen very quickly. But the people living in Europe had been there for 25,000 years before the beginning of farming. What could have been an engine that drove a widespread change in Europe? There was migration into Europe by the Linear Bandkamerik (LBK) from the East for about five hundred years beginning in about 7500 BP, which apparently is advocated by some as a cause for the change in skin color. Europe was already settled and thriving for 25,000 years and the whole subject of diet and vitamin D was probably superfluous for them whether the influx came from the Mid-East or the Baltic area.

Let us ignore this weakness in the theory of Sweet; let us instead see about his hypothesis, that to knock the theory out, one must show evidence that Europeans, as early as the Magdalenian cave art, were as fair as modern Europeans, or in other words, at least over 10,000 years ago. How the five millennia turned into 10,000 years, is a mystery but we will argue on his terms.

And, in short, Sweet argues that one can disprove his theory of the later date only by proving the correctness of the other, of an earlier date. This is not very helpful. It could as well be said that the validity of the earlier date could be disproved only by proof of the validity of the later one. Sweet probably feels comfortable relying on the brown paint used to draw the stick figures of the hunters in some cave art, confident as he must be of no other way to prove the earlier date.

The fact that other civilizations lived in zones of insufficient sunlight and had diets insufficient in vitamin D, yet did not grow pale like the Europeans does not, for him, disprove his theory. He explains the continuing dark skin, tan or bronze, but not pale, as the populations in question still ate meat. The fact is that both Europeans and mid-easterners also continued to eat meat, yet paleness did come. There are vegetarians in the world, but they are individuals, not entire cultures. Some cultures may not eat beef; some others may disdain something else. But what culture refuses all beef, fowl and fish? Eating meat is universal and has so been for humans since humans have been here.

As if evidence were needed to prove what is so generally known, a recent study provided scientific evidence challenging any notion that farming immediately and completely changed everything about human society. Much, including diets, can be inferred from analysis of stable isotopes in human bones found at archeological sites in Europe. They found that hunting and gathering persisted even after farming had been established. Their conclusion is based mainly from the presence of specific lipid biomarkers left inside ceramic vessels that date from the time that plants and animals were first domesticated.[142]

Apart from the failure to prove any formula that would cause pale skin, we still have the question of the timeline for the change that obviously did occur. There is another way to account for the earlier date, but, though I have never met the gentleman. I feel certain it is one Sweet would not accept. One simply does not mention UFOs or extraterrestrials in polite society.

And the idea does very much involve extraterrestrials. They may be the only people on Earth, and there are some on Earth, who have skin like that of the Caucasians. They have undoubtedly been visiting, and perhaps living on our planet, for many millennia, possibly even before there were any anatomically modern humans, possibly before any humans, almost four million years ago. It is possible that through in vitro fertilization, and later, perhaps also through intermarriage, or, in any event, sexual relationships, they have guided our destiny, and our genes, to what we are today. By most scientists this possibility would not be considered.

But let us answer Sweet on his own terms, namely the requirement that we prove pale skin before the end of the time of the cave art, which is the Magdelenian. I have mentioned one exception to the practice of depicting humans as stick figures. Sweet has apparently overlooked it. It is only one of a number of events connected to the great transition, but at this point it needs to be front and center.

PART TWO

SO SUDDENLY HUMAN

CHAPTER XII

MIRACLE AT LA MARCHE

WHAT IS THE ONE EXCEPTION previously mentioned, to the otherwise universal style of stick drawings of early humans? If we are inclined to exaggerate the suddenness of the change in skin color, we may need to consider that exception. In fact, we should consider it in any event. It does not take place at the very beginning of the appearance of fully modern humans, but it is one of vital importance to us.

The exception is a 120 square meter rock shelter known as La Marche. It is near the French town of Lussac-les-Chateaux, about 75 miles south of Tours, located in a valley near a stream. Archaeologists believe it was occupied only seasonally and probably only during the day. A treasure has been found there, consisting of over 1200 engraved slabs of stone, varying from a few centimeters to over a square meter weighing over 40 pounds.[143] The slabs came from further up the valley, as the ones indigenous to the shelter were not suitable for engraving. The engraved slabs are of much finer quality.[144]

Among the species of animals depicted on the slabs are about 115 of our own species, including 51 practically complete humans with heads and bodies, and 57 isolated heads. Of those that can be identified by sex,

13 are male, 27 are female.¹⁴⁵ What makes the site and the engravings so unique is the fact that many of the human faces are recognizable humans. They have individualized features and characteristics, so much so as to seem like personal portraits. They may be our only opportunity to see the cave people as human beings.

Because of heavy overengraving, it is not easy to make out the individual faces. We are fortunate that the task of doing so has been accomplished by Dr. León Pales. It must have been painstaking work to first identify, then to extract from the morass of lines the individual portrait. We can get some sense of that difficulty by viewing the bearded face from one of the blocks and entanglement of lines from which it has been extracted (ill 30). Twenty of the heads that were extracted and modified from the original to compare on the same scale and facing the same direction are shown in ill 30. The work has usually been dated to about 11,000 or 12,000 BP. But other "guesstimates" run as high as 15,000 BP.

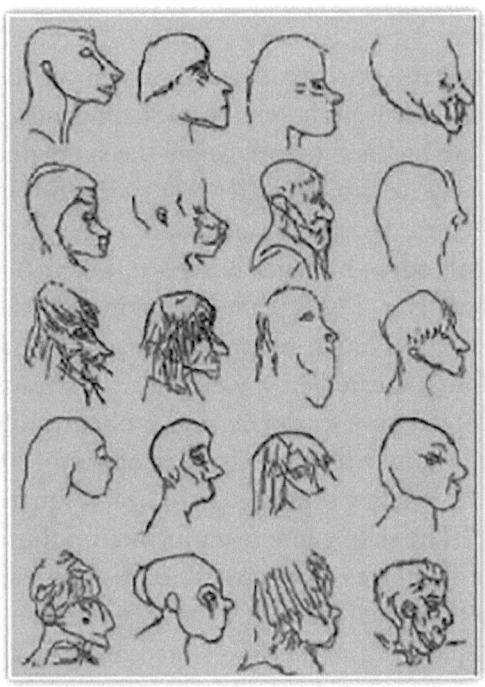

Illustration 28: Twenty Faces from LA Marche.
Credit: http://news.bbc.co.uk/go/em/fr/-2/lu/science/nature/2012385.stm

Illustration 29: Profile of face of old man from La Marche and stone from which it was extracted.

Credit: See Ill 28.

Illustration 30: Frontal of Old Man from La Marche, and stone from which it was extracted.

Credit: Guthrie and see Ill 28.

Illustration 31: Three portraits of humans are engravings on stone slabs at La Marche, Vienne, France, more than 14000 years old.

Credit: P. Bahn 'Prehistoric Art'

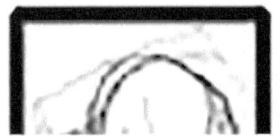

Illustration 32: Three portraits from La Marche. On the right: Grotte de La Marche, Lussac-les-Chateaux (Vienne, France). Faces of two children wearing what appear to be hats. The face on the left is the same as in 30.

How does this relate to our current subject of skin color? It is speculation but hopefully informed speculation. The color of the faces on the slabs is white, which means nothing to us. That is simply the color of the slab. Limestone is always white. But the change in skin color did not happen overnight, nor did the difference in facial features. We might assume however that they have changed roughly in tandem. Viewing these faces is to me, as the eye of one beholder, a study in gradualism.

At that late date, almost the end of the Upper Paleolithic, I would have expected to see all Caucasian facial features, but it seems to be a mix, mostly Caucasian, but a mix nonetheless. I realize that too much can be made of physical "racial" characteristics, but in classification of fossils they are still universally used. In modern times such characteristics mean very little. There has been not only much intermarriage, but changing life styles have no doubt changed many of the physical markers by which we determine Caucasian, Negroid or Mongoloid features.

Those markers however, after all, did evolve for a reason, and could disappear or become less pronounced when the reasons disappear. But we are dealing with portraits done in prehistoric times when the original population characteristics were probably much more pronounced. Comparisons are best not made with individuals. The bigger the sample the more significant are the findings, or lack of findings. There is undoubtedly overlap not only between members of different populations, but also between members of the same populations.

This is a very small group, only 26 faces (There are a few duplicates that I trust I have not counted as 2), but still a group, not simply one individual, and it may be worthwhile to consider whatever evidentiary value it may have. What are the markers, the characteristics, we look for? According to the Collins English Dictionary. "Negroid" denotes "one of the major racial groups of mankind, characterized by brown-black skin, tightly-curled hair, a short nose, and full lips. This group includes the indigenous peoples of Africa south of the Sahara, and their descendants elsewhere."

But this dictionary definition lacks a number of other characteristics. These include a broad and round nasal cavity; "Quonset hut-shaped" nasal bones; notable facial projection in the jaw and mouth area, known as "prognathism"; a square or rectangular eye orbit shape;[146] and large teeth. It is the prognathism, the elongation of the face that is the most readily observed and is indeed most readily observed in some of the profiles we see from La Marche. It has also been reported in scientific journals that whites generally have more facial hair than blacks, which I believe is a common observation. In any event the hair on the portraits does not seem tightly curled; it seems to flow, more often the case of the hair of Caucasians.

In a more comprehensive summary, created in 2011,[147] one completely adequate for our purposes, characteristics of the skulls of Caucasians and Negroids are contrasted.[148] Obviously the physical traits of Caucasoid crania are still recognized as distinct (in contrast to Mongoloid and Negroid races) within modern forensic anthropology. A Caucasoid skull has often been identified by archaeologists and anatomists with an accuracy of up to 95%, by the following features:

Little or no prognathism exhibited—an orthognathic profile, with minimal protrusion of the lower face (though the term means that the lower jaw neither projects nor recedes from the upper); Retreating zygomatic bones (cheekbones), making the face look more "pointed"; Narrow nasal aperture, with a tear-shaped nasal cavity.

More specifically, heads of Caucasoids are said to be long, with a width of less than 76 % of the length, a type known to scientists as "dolichocephalic." They generally have little supraorbital development, meaning the bone structure around the eyes tend to be rounded. The face also tends to be narrow, a characteristic known as "leptoprosopic," meaning having a long, a narrow, or a long and narrow face. This term is a measurement of height, not length, namely the height from the top of the interior aperture of the nose to the chin, divided by the width between the cheekbones, the result being known as the facial index. A facial index of 88.0 to 92.9 as measured on the living head and of 90.0 to 94.9 on the skull is deemed high, or leptoprosopic.

There is most often neither facial nor "alveolar" prognathism in Caucasoids, the quoted word referring to the ridge of the gums behind the upper front teeth. In short, the front teeth do not protrude. However the authors state that it sometimes occurs in Caucasians among some archaic peoples, who are precisely the ones we may be dealing with here. The noses are long, narrow, and high in both root and bridge.

The heads of the Negroids are described as usually like the Caucasians, Dolicephalic. A small minority however are brachycephalic, meaning a relatively short or broad head; more specifically having a head whose width is 81 % or more of its length. The forehead is most often high. There is little supraorbital (bones around the eyes) development. The face is leptoprosopic to a much lesser degree than that of the Caucasian. Prognathism is common in most Negroid populations. The nose is low & broad in root and bridge with depression at the root.

Another summary of Caucasoid features, compiled in 2009[149] is similar. Unfortunately the summary of the Negroid features is limited to those living in American or European areas, which would be of no value to us.

It describes the Caucasians as the northern population, living in the northwestern Europe, namely, Scandinavia, Iceland, Britain, Denmark, northern Germany, and "elsewhere," as having the following characteristics:

> Skin color is white or pink, some are freckled faced. The face is narrow and high. There is no visible prognathism. Eyebrows are slightly raised. The hair is blond, and from light gray to pale chestnut. The forehead is high and slightly receding. The nose is thin, with narrow nostrils and nasal bridge. Some have straight, aquiline noses. The eyes are small, gray-blue or green. The lips are thin and the oral fissure is narrow. The chin is well defined and the lower jaw is high.

All of the authors of the writings I have cited say emphatically the differences are superficial and result from adaptations over many centuries or millennia to their environment. They stress that there are differences, but that the differences are unimportant. We are all one species. The one importance is in identifying fossils or for forensic, legal, purposes in identifying unknown corpses for instance. But that is precisely why they are, or might be, important to us.

With these features in mind, what do the profiles from La Marche have to tell us, if anything? Only one of the faces is a frontal view, namely the one extracted from the entanglement of lines in ill 8c. He has something of a beard, and certainly not the full lips of the Negroids, many of whom have this feature even today. From the front we can tell nothing about an elongation of the face; the nose does not seem flat or wide. But many of the 20 profiles show evidence of prognathism, the elongation of the face, including three on the left of the top row, and the two left faces on the second row. Many of the group seem to have much hair, none of which is tightly curled. In the third row, the third from left is an interesting study. He is the best candidate for the short flat nose of the Negroid, and I would not know what to make, if anything, of what appears to be a double chin. I have assumed most of the subjects to be male. They look it. But then I really don't know how the faces of men or women looked 12,000 years ago.

The opinion of any reader or observer of the illustrations will be at least as valid as mine. I sense that these persons are considerably more Caucasoid than Negroid, hence probably more white than black, but I think also that it is a matter for the eye of the beholder. I would not be shocked to learn that they are all Negroid, or all Caucasian. I have omitted any mention of the Mongoloids, the Orientals, not for lack of interest, but because they play no role in this particular aspect of the history.

But whether the engravers were white or black. Caucasoid or Negroid, they were artistic geniuses of the highest order, probably working entirely without training or guidance from any other contemporaries. They were keen observers also of the faces of their fellow clanspeople, something apparently unique in the Paleolithic. The Egyptians, for all their technical skills, never painted individuals, only types, sometimes long rows of people who look exactly the same, without individual traits, any indicia of character, or facial expression.

Sympathetic critics call that almost universal "Egyptian" style "idealized., "generalized, or "stylized." Other critics, Jean Cocteau, for one, see little that is admirable in them and is especially disdainful of the term "stylized." He said it was a word "invented by the witty to designate things that absolutely lack style." Not until the advent of the Ancient Greeks, over 10,000 years later, do we again see the likes of these individualized portraits.

CHAPTER XIII

TRANSITION IN AFRICA AND EUROPE

THE ANATOMICALLY MODERN SPECIES OF Homo sapiens was preceded by other species known as Archaic Homo sapiens beginning about 500,000 BP. It was about this time also that the last common ancestor of both anatomically modern humans and the Neanderthals lived. Homo erectus, disappeared about 300,000 BP, leaving their descendants. Homo ergaster and Homo heidelbergensis. Anatomically modern humans evolved in Africa sometime over 100,000 years later.

The first migrations out of Africa by earlier members of the genus Homo were sometime after about two million BP. Much later, but no later than about 100,000 BP anatomically modern humans spread out, first, into Arabia, then to the Middle East, followed by many parts of Southeast Asia. They had reached China, about at least, 800,000 BP. The earliest evidence of any members of Homo sapiens in Europe, at a single site. Atapuerca, Spain, appears to be of the same approximate date. However, as of this writing, there is no further definitive evidence of human occupation in Europe dated before 500,000 BP.[150]

Anatomically modern Homo sapiens of the Middle Paleolithic (MP), the archeological period preceding the UP, are our most recent predecessors before the advent of fully modern humans, called (by ourselves) Homo sapiens sapiens. Yet the anatomically moderns showed little difference in behavior or in intellect from earlier species, or their contemporaries, the Neanderthals, whose culture was known as "Mousterian." Neanderthals inhabited Europe, the Middle East, and western Russia from about 250,000 until about 30,000 BP.

There were certain advances by both cultures over their common forebears. It might be instructive to look at those ancestors, most likely our earlier predecessors, namely, Homo erectus. The erectus culture was called the Acheulean, and was known primarily for its hand axes. Their origins date to about 1.8 million years ago. Slightly older fossilized remains of erectus have been found. It was they who made the first journeys out of Africa. Remains of fossils or tools, or both have been discovered in Europe, Africa, the Middle East and the Far East.

They survived until about 300,000 BP. though there are disputed findings of one or more younger fossils, dated to about 60,000 years ago.[151] About half way through their Earthly existence they began shaping bifaced tools rather than those with a single edge. During their last 600,000 years their tools became more symmetrical in shape, hence more aesthetically pleasing. The later tools however were no more utilitarian than the earlier ones. The shaped stones were essentially the same, fashioned for killing animals, and butchering or skinning, or pounding or breaking of foods.

Two archaeologists who spent their entire careers studying them found the almost changeless continuity over millions of square miles for a million and a half years, "absolutely astounding."[152] Whether we are to any extent descendants of erectus or of Homo ergaster, or heidelbergensis, is not certain. In some respects ergaster had physical and possibly behavioral qualities closer to humans than did erectus, but to an almost certainty their tools were similar to those of erectus and persisted with little change for a similar period.[153]

Neanderthals existed from sometime after 500,000 until about 30,000 BP. There are minor anatomical differences from modern humans, including some peculiarities of the shoulder blade, shape of the skull, dentition, and of the pubic bone in the pelvis. Neanderthals mostly lived in cold climates, and their body proportions are short and solid, with short limbs. Men averaged about 5'6" in height. Their bones are thick and heavy, with apparently powerful muscle attachments. They were formidable hunters, and are the first people known to have buried their dead, the oldest known burial site being about 100,000 years old. They are found throughout Europe and the Middle East.

The last of the Neanderthals died about 30,000 years ago, just 10,000 years after the influx into Europe of Late Stone Age humans from Africa, with whom they could not compete for the same resources. Or, as said by linguist Christine Kenneally,[154] we afforded them too much competition or too much loving by interbreeding with them and eventually swamping them with our far larger population. Or, she also suggests that Homo sapiens may have hastened their end by bringing new diseases into the Neanderthal world.

The Neanderthal and modern human lineages have been genetically determined to have split about 500,000 BP, the date of their last common ancestor, though there is little or no fossil evidence from that age. Genetic evidence further indicates probably no significant interbreeding. Hence, it appears that the Neanderthals have contributed little to the lineage of present humans.[155] The most recent word on that subject is to the effect, that humans outside Africa carry about 2.5% Neanderthal DNA. It is also reported that people from parts of Oceana carry about 5% DNA from a population living in Asia that also probably became extinct about 30,000 years ago called Denisovian.[156]

These earlier populations, including anatomically modern humans may have been the closest relatives to us from prehistory, but they are obviously very distant relatives. All other groups, even those similar to fully modern humans in Africa in body structure, survived until about 50,000 BP when they succumbed eventually to fully modern humans immigrating into Europe. Before looking at the changes taking place in the transition, we should glance at the lifestyles and habits of the anatomically modern humans, which are similar to some of the other earlier groups.

These earlier populations struck stone flakes from prepared cores; built fires and buried their dead, but without grave goods or other evidence of ritual. They sometimes acquired large mammals as food. Like the Neanderthals, and like their own earlier ancestors, they manufactured only as small variety of stone tool types. They obtained stone mostly from nearby sites, indicative of little ability to explore or organize; they rarely used bone ivory or other similar materials to produce artifacts. Nor is there other evidence of structures at camp sites or elsewhere. They were not efficient hunters, and apparently, with few exceptions, lacked the ability to fish. Populations were low and there was no indication of art or bodily or other decorations.

But within a few thousand years, beginning about 50,000 BP, dramatic changes occurred. Ofer Bar-Yosef, in 2002 summarized many of them:[157]

As opposed to flake production in the Middle Paleolithic (MP), UP assemblages usually involve systematic production of bladlets, sometimes shaped into microlithic tools of various forms, and of prismatic blades, that is, those whose ends are parallel, with four or more sides and angles that are parallel and equal.

There were relatively rapid shifts within several centuries, or a few thousand years at most, in core reduction strategies as well as bone and antler tool design. These shifts are interpreted as reflecting changes in style and transmission of cultural information and rarely are related to functional needs

There was exploitation of bone and antler as raw materials for the production of daily or ritual tools and objects. This became a common practice in the UP. Though these raw materials were common in MP sites they were generally not exploited.

There was systematic use of grinding and pounding stone tools, something that began during the UP. This is best documented where plant food played a major role in the diet such as in the Mediterranean region and Africa. None of these tools were found in MP contexts, although there was consumption of vegetal substances during that earlier period.

Systematic use of body decorations—beads and pendants—made from marine shells, teeth, ivory, and ostrich egg shells are recorded from both Europe and the Levant, that is, lands bordering the eastern Mediterranean. These are considered to communicate the self-awareness and identity of both the individual and the social group. No similar objects, or other clear signs for the identity of the social units, were found in Middle Paleolithic contexts.

During the UP, for the first time there was long-distance exchange networks in lithics, raw materials, and marine shells reach as far as several hundred kilometers. This consistently differs from the much shorter ranges of raw material procurement during the MP

The UP witnessed the invention of improved hunting tools such as spear throwers, and later bows and arrows and boomerangs. They devices facilitated targeting animals from longer distances and probably brought higher rates of hunting success.

Human and animal figurines, decorated and carved bone, antler, ivory and stone objects, and representational abstract and realistic images, painted or engraved, began to appear in caves, rock shelters, and exposed rocky surfaces by 36,000 years ago.

Bar-Josef also points to intrasite features including burials, and subsistence strategies. These include Storage facilities, generally known from northern latitudes where underground freezing kept food edible. Storage occurs in UP sites after the initial phase. None of these structures were recorded in MP contexts. Structured hearths with or without the use of rocks for warmth banking and parching activities were recorded in Upper Paleolithic sites. Variable types of hearths are known from both MP and UP periods, although the use of rocks is almost exclusively documented from contexts of the latter period.

Also mentioned by Bar-Josef are distinct functional spatial organization within habitations and hunting stations, such as kitchen areas, butchering space, sleeping grounds, discard zones, and the like, which are relatively common in UP sites. Such features, though better preserved in the later phases, about 20,000 BP, even the very early UP sites produced good examples. It all seems to point to a changed perception of space. According to Bar-Josef, such features may reflect the social structure or a particular combination of members of the band, such as a male task group. This kind of evidence is rarely found in MP sites.

Possible differences in subsistence activities can also be taken into account as differentiating the Middle from the Upper Paleolithic. As knowledge concerning the exploitation of plants is poor, most of the evidence centers on the issue of hunting versus scavenging. The evidence clearly demonstrates that both Middle Paleolithic and Upper Paleolithic humans were hunters.[158]

However, it appears that only Late Stone Age people, a term used for Africa which is the same as the UP, particularly in southern Africa, routinely fished and fowled, as evidenced by many more bones of fish and airborne birds at those cave sites that date from the Late, as opposed to the Middle Stone Age. With regard to mammals, there is a dominance of bones of elands from the MSA, an animal that was much less dangerous to hunt than buffalo or wild pigs, whose bones predominate after 50,000 years ago. This has been deemed to reflect advances in projectile manufacture, including, after about 20,000 BP, the use of bows and arrows. These new hunting targets appear also related to the increasing population of the LSA.[159]

Not all archaeologists agree that the change was so sudden, or that it resulted from genetic change. They claim instead that it was a gradual and cultural change from the Middle Stone Age. Some point, for instance, to other areas of the world where certain advances coincided with or preceded similar changes in Africa or Europe. Practically all such changes however are in tools and weapons; in extremely few are there examples of any of the other changes that signal a change in new perceptions of time or space that are the ultimate mark of the sudden African transition and the accelerated continuation of it in Europe.[160]

There may indeed be grounds for legitimate dispute on many of the aspects of that revolution, and differing viewpoints of the issue between sudden versus gradual, and between genetic change in the function of the mind versus cultural development with the same mental equipment. But Bar-Josef found the difference in the art of western Europe and Australia puzzling.

One must wonder, he says, why western Europe and, in particular, the Franco-Cantabrian region is so different from the rest of the Upper Paleolithic world. It is not the lack of limestone caves or suitable rock surfaces that prevented other social groups or their shamans from leaving behind similar paintings and engravings. Possibly this local flourish had to do with the vagaries and pressures faced by foragers in two major isolated regions at the ends of the inhabited world—Western Europe and Australia—where there are claims for rock art of the same general age. If this explanation has any foundation, he concludes, we should look for the details of the common behavioral denominators.

But the subject is of more significance than is accorded here by Bar-Josef and deserving of more extensive treatment. No one has ever satisfactorily explained genius. We will take a closer look at those paintings in the next chapter for our own purposes, but without any attempt to explain them.

CHAPTER XIV

OF SPACE AND TIME: THE BEGINNING OF ARTISTIC GENIUS, AND CALENDARS

THERE ARE TWO MATTERS IN the summary of Bar-Josef that almost shout genetics, one of which will brook no dispute. They both involve fundamental attributes of the human mind making quantum leaps in the Upper Paleo-lithic, namely, the sensitivity to space and to time. In addition to the division of space for different living purposes, there was the first appearance of art. For the first time we see also what are, in all probability, calendars.

Art appeared in many areas of the world, and in at least two areas, Africa and Australia, perhaps as early as in Europe, if not earlier. These may mark the beginning of art, but only in Europe do we see the beginning of artistic genius. It was this difference that caused bafflement to Bar-Josef. It came in western Europe like a stroke of lightning; almost literally like the mythical Athena who sprang fully formed from the brow of Zeus.

The earliest of these marvels of paintings and engravings are contained in Chauvet Cave in south central France, and Cosquer Cave, now partially submerged under the French Mediterranean. The earliest paintings in Chauvet date from about 35,000 years ago; Cosquer, a few thousand years later. Deep within Chauvet, Cosquer, and the more recent caves the works are remarkably realistic depictions of animals: horses, bison, bulls, reindeer, lions, panthers, bears, owls, hyenas, and rhinos among others. In some discovered recently, in addition to bison, ibex, and horses, are also marine animals such as seals and what appear to be jellyfish and diving birds known as auks.

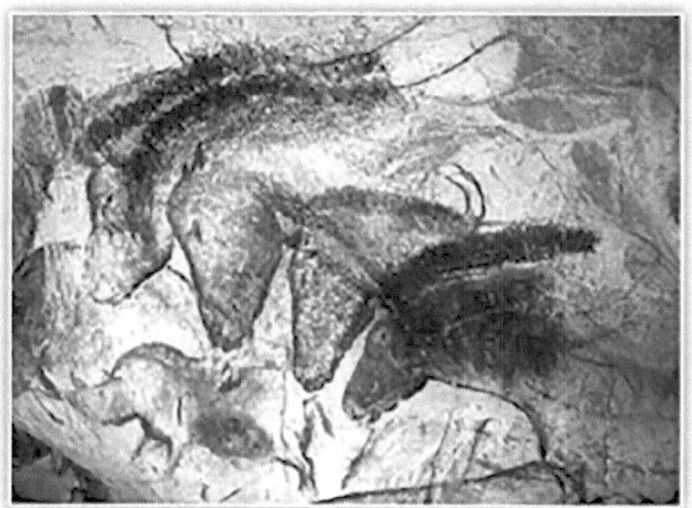

Illustration 33: Hoses painted in the Chauvet Cave, in the Ardèche département, southern France. It was first explored in 1994. It contained the fossilized remains of many animals, including those that are now extinct, and the floor preserved the footprints of animals and humans.

Illustration 34: Cosquer Cave, earliest paintings dated to about 30,000 years ago. It was discovered in 1985 by a diver, Henri Cosquer. The original entrance is about 115 feet below present-day level of the Mediterranean.

Similar works have been found in other caves throughout France, northern Spain and western Germany, mostly dated to approximately 18,000 to 12,000 years ago. In those caves with the more recent paintings however are sometimes sculptures that date from a time closer to the paintings of Chauvet. There are, for example, in early layers of several caves in south-west Germany seventeen beautifully carved ivory figurines that are as old, or older, than the paintings at Chauvet.

Among the best known to the general public are the caves of Lascaux in France and Altamira, in Spain, but there are over 350 of these European sites. They contain over 15,000 paintings and engravings,[161] and over 10,000 items of portable art, namely sculptures.[162]

On every continent except Antarctica there are tens of thousands of prehistoric paintings, up to 175,000 in South Africa alone. Though they may be more recent, from only 10 to 12 thousand years ago, and some only centuries old, others in Namibia, Zimbabwe, India, and Australia may be from 40,000 years ago or older.[163] What makes these European works unique is neither their numbers nor their age, but the brilliance of their artistic quality.

Ann Sieveking, for one, says that in comparison with the European works, Australian prehistoric art is unsophisticated and crude.[164] Another archaeologist says that they compare favorably with any modern ones. Still another declares that these animal figures are works that no later culture, including civilized ones, could surpass.[165] Perhaps the most perceptive comment comes from Arnold Hauser, who marveled at the mastery of foreshortenings.[166]

That term refers to three dimensional perspective as applied to representations of individual objects on a flat surface. It means, necessarily, an appreciation of the beauty in depth, the third spatial dimension. Except for a very short episode in Egypt around 1200 B.C., for which there was little if any appreciation or understanding, we see no more of foreshortening until the time of the ancient Greeks, over 10,000 years after the last of the cave art. It largely disappeared in the Middle Ages, when again there was little popular aesthetic appreciation of it. But it returned in full bloom in the European Renaissance.

Most of the cave animals however are not foreshortened, but are shown in a twisted perspective, every part being shown to best advantage, no matter whether seen in such view in real life or not. This style is typical universally in ancient times, except for the Greeks and Romans. It is typical also in more recent art cultures, such as those of the Maya, Aztec, Inca and North American Indians until acculturation by contact with those who followed Greek forms.

Interestingly there are relatively few depictions of humans in the European caves, and those few are often "stick figures" with lines for arms and legs, circles for faces, lines for facial features, if any, and boxes or lines for bodies. Further, there are no real scenes, or depictions of nature: no rivers or streams, mountains or hills, flowers, vegetation or open fields. There are some geometric forms: squares, circles, spirals, ellipses, tent shapes, wing shapes, checkerboards, tridents, stars, and curvilinear mazes among others. But about 95% of all works are animals.[167] None of this can detract from the awesome animal displays on the caves' walls. They have mesmerized countless researchers and visitors for many decades.

Many believe the change in manifestations of the human brain between the Middle and Upper Paleolithic were biologically determined by changes in the brain's modularity structure, resulting, for the first time, in a sharing of information within the brain concerning nature, social interaction and technology, all fields of information being coordinated.

Bar-Josef has written that this could only have happened if the additional neurological change in the brain took place as suggested by Klein and Edward.

The observation of Klein and Edward: "We suggest that this capacity stemmed from a genetic change that promoted the fully modern brain in Africa around 50,000 years ago."[168] The period during which the artworks were created in Chauvet was known as the "Aurignacian," which Klein and Edward place between 37,000 and 29,000 years ago, and as stretching across southern Europe from Bulgaria to Spain.[169]

Genetic it must have been. Klein and Edgar also emphasize the stark drama of the change, in all its manifestations from the Upper Paleolithic from that of the Neanderthals, whom the African immigrants quickly replaced. They speak of the remarkable monotony of the Neanderthal culture over tens of thousands of years in comparison with the rapid diversification in the work of Homo sapiens sapiens that occurred from the Aurignacian period onward.

Nicholas Wade notes that in the initial years of the immigration from Africa following the initial stage of the transition, people everywhere must have looked much the same. Despite that sameness, however regional differences became inevitable. For archaeologists, he continues, "The most striking are artistic. There is nothing to match the great painted caves of Europe." He cites the words of Bar-Josef we have seen above, and comments that historians and social scientists tend to offer purely cultural or environmental explanations for all differences. From a biologist's view, however, he states that it seems likely that genetic influences must also have been at work.

He prefaces his conclusion with this explanatory matter:

> There was a significant difference, or the seeds of a difference between the European and Australian antipodes of the modern human advance from Africa. The Australian and New Guinean branch soon settled into a time warp of perpetual stagnation. They were still living with Paleolithic technology when their European cousins came visiting 45,000 years later. They never broke free from the triple

bonds of patrilocal society, nomadic mobility and tribal aggression. For some reason, the modern people who reached Europe and the Far East were able to escape this trap and to enter on a phase of steady and continued innovation.[170]

For what reason were these differences to develop, and why in such a pronounced manner? Wade buttresses his belief in the genetic element with some interesting facts.[171] He cites a discovery by Bruce Lahn, a geneticist, involving two genes, each of which had alleles (forms) that conferred a cognitive advantage in the human brain. The advantages were slight, but sufficient to cause evolutionary selection. One of the alleles is a version of a gene known as microcephalin, which first appeared about 37,000 years ago, about the time of the origin of the Aurignacian and perhaps only 2000 years before the first paintings in Chauvet cave. Dating is not precise enough however to exclude a large room for error, and the period of its possible beginning of the allele extends from 60,000 to 14,000 years ago. It is now carried by about 70% of many populations in Europe and east Asia. In sub-Saharan Africa however it is generally carried from zero to 25%.

The other allele is a form of another brain gene known as ASPM, and appears in Europe and the Middle East and is carried by about 50% of the populations there. The allele is less common in East Asia, and almost totally absent in sub-Saharan Africa. Interestingly, the time and place of the appearance of this allele coincides relatively neatly, according to some sources, with the beginning of agriculture, the second transition referred to earlier in this space, and likewise, accordance to some, with the alleged first occurrence of "white" skin.

Still, that other questions of "why" rears its head. What precipitated these changes only in the Europeans and perhaps in populations of the Middle East? Wade limits his reasons to those limited to the traditional ones, reactions to local conditions, evolutionary pressures, and to random genetic drift. Of those issues, only genetic drift would seem to apply to the focus of our attention here. It is very difficult to understand what evolutionary pressures could have been at work that required splendid works of art, or what local conditions could have contributed.

Both Bar-Josef and Wade addressed the issue of possible influence by differences in local conditions and rejected any likelihood of it. This concerns a strain of genius that ran through a population for about 25,000 years, one that knew no competitors, nor any practical necessity. For other causes, perhaps we should turn to other scholars, namely evolutionary geneticists.

Before doing so however we should be aware of other changes during that period, one of which, crude, but probably accurate, involved calendars based on phases of the Moon. The other is of necessity, mostly speculation, of which there has been plenty. That involves uses of language in the modern sense, with logical structure and ability to convey complex thoughts.

We turn first to the subject of the calendars. It shows a perception of time, and sensitivity to it.

* * *

In 1963 Alexander Marshack became a research associate at the Peabody Museum of Archaeology and Ethnology at Harvard University. In the course of his employment he gained access to state and university archaeological collections. In 1972 he rose to public prominence with the publication of *The Roots of Civilization*.[172] The most startling of his proposals was the theory that notches and lines carved on certain Upper Paleolithic bone plaques were in fact notation systems, specifically lunar calendars notating the passage of time.

Using microscopic analysis, he demonstrated that seemingly random notches on bone were sometimes actually a structured series of numbers. Most significantly, he claimed that notches, or pitting, on a bone plaque from the Grotte de Thaïs in southern France, which dates to approximately 12,000 BP, were structured in subsets of 29. He hypothesized that they were used to mark the duration between one new moon and the next, a lunar month.

Another similar find was a bone plaque from the Abri Blanchard, Sergeac, in the Dordogne. According to Marshack, there were sixty-nine marks with twenty-four changes in the type of pitting, (see Illustration 35 below). The type of technique changes with the different phases of the moon, crescent-shaped, full or dark. The Abri Blanchard plaque bore eighty-one marginal marks which, in addition to the original sixty-nine, would comprise a record of about six months.[173] The proposition was debated by a number of archaeologists who claimed he was reading too much into the markings.

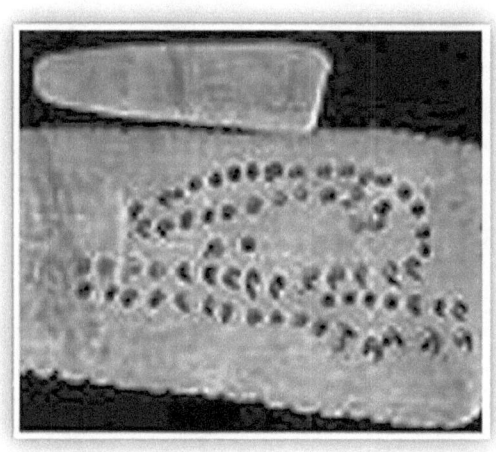

Illustration 35: Bone plaque from the Abri Blanchard, Sergeac, France, with enlargement of the series of pits.

Much of the steam was taken out of the arguments of his opponents with the discovery in 2000 of markings on the walls of the famous cave of Lascaux, likewise in southern France.[174] The markings and the paintings nearby are dated to about 15,000 BP, about three thousand years older than the stone notches in the Grotte de Thaïs. They were interpreted by Dr. Michael Rappenglück of the University of Munich. He claims that groups of dots and squares among some of the animals depict the 29 day cycle of the Moon. If he is correct, as seems likely, it is, so far as known, the earliest calendar ever found. He says that the paintings on the cave walls by Cro-Magnon man were done during a time of temperate weather in the Dordogne Valley, where Lascaux is located, though much of Europe was in the grip of a period of glaciation with its freezing temperatures.

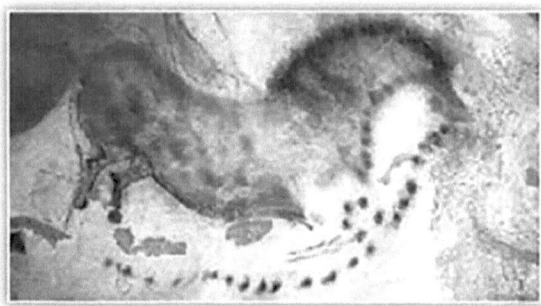

Illustration 36: What could be the oldest lunar calendar ever created has been identified on the walls of the famous, prehistoric caves at Lascaux in France.

Credit: BBC News Online science editor Dr. David Whitehouse.

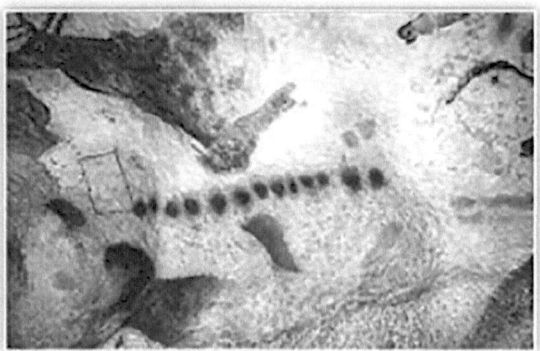

Illustration: 37: Half Moon cycle, 13 dots and an empty square

Below a painting of a deer, as shown above, was a row of 13 dots, ending in a square. According to Dr. Rappenglück that represented half of the Moon's monthly cycle, one dot for each day the Moon is in the sky. At the new Moon, when it vanishes from the sky, he continues, the empty square "perhaps symbolically represents the absent Moon." Further along he points out, beneath a dappled brown horse with a black mane there is a row of dots. They total 29 in all, "one for each day of the Moon's 29-day cycle, when it runs through its phases in the sky." He explains another series of dots that curve away from the main row. "I think that indicates the time of the new Moon, when it disappears from the sky for several days."

PALE SKIN, GIANTS, AND THE GREAT TRANSITION

Illustration 38: The Pleiades star cluster sits above the bull's shoulder.

For almost two million years of human existence, there has been, before these discoveries, no evidence of time keeping or any concept of the passage of time. Such a concept requires a feel for sequence, what should follow what. That is the way we talk, the way we make sentences. Both depend on our inborn sense of logic, of order. This brings us to the subject of language, the biggest mystery of all.

CHAPTER XV

LINGUISTICS

Paintings and calendars can be seen and examined and thanks to modern science, can often be dated with reasonable accuracy. With oral sounds, of whatever kind, there is no such possibility. The last echoes of the first sounds of humans have long since disappeared, never to be recovered. Nor has any later speech survived until the beginning of writing, probably no earlier than about 5500 BP. The history of investigation into the subject of the beginnings of human speech has been one of speculation, ranging from the frivolous to the convincingly analytic.

There are many components to speech, but only three major aspects with which we need concern ourselves for possible light on our subject: vocabulary, syntax, and pragmatics. With that last category we are concerned only peripherally and after a few observations, we can leave it to focus on the first two.

Pragmatics does not rear its highly interesting head until the time of the Greek poets and dramatists, long after written records accumulated, and we know when and where, and perhaps why the use of pragmatics arose. It is worth mentioning for what it may tell us of the relatively sudden growth of the Greek culture, which, in turn, may give us a glimpse into the beginnings of human speech, tentative though it may be.

Pragmatics encompasses metaphor, the use of one word, phrase, sentence or paragraph, sometimes an entire story, as a symbol for something else. We might talk of our journey through this vale of tears, for living a sad life; we might call roadside signs warts on the face of the land. Pragmatics is the understanding of what is implied as opposed to what is literally said; reading between the lines as opposed to what is in them. It appeals to the imagination and there is an element of pleasure in making the leap from the spoken line, the symbol, to what it implies. It first arose in Ancient Greek poetry and particularly in the great tragedies. It is a quality lacking in the literatures of the ancient Egyptians, and all other contemporary cultures, including the earlier Greek. The epics of Homer are filled with the precursors of metaphor, namely similes, but the few metaphors are used often enough to become tropes, not true metaphors.

There is also in the classical Greek literature, the obvious need for novelty, and creativity, as opposed to the repetition, often deadly, of words and phrases in all ancient writings before the Greek, the Old Testament included.

The appearance of novelty and metaphoric language arose in the Greek culture within the same time frame as three dimensional perspective, facial expression and individuality in portraiture, the study of "man" in their philosophies, the recognition of the worth of the individual, the rise of a true democracy, and the beginnings of the love of nature for its own sake. All were attributes missing either entirely or in significant part in other contemporary cultures.

Steven Pinker, examining the possibilities for the beginning of spoken language[175], denigrates archaeologists who, he says, are largely unaware of psycholinguistics, to "pin language" to the same time as the "cultural fads of the Upper Paleolithic." Such conclusions, he says, depend on there being a "single symbolic capacity underlying art, religion, decorated tools, and language, which we now know is false." Pinker probably means by the "single symbolic capacity," a single gene. If so, he is probably on good ground in dismissing that possibility. However for the most part, all of those new attributes are, by our own contemporary psychologists, psychiatrists and others, attributed to the right hemisphere of the human brain. That includes, most particularly, facial expression in art, pragmatics in speech and understanding of it, and interest in human nature and individuality.

It seems just as likely that the bundle of new attributes about 50,000 years ago, demonstrating a vastly increased sense of logic, orderliness, planning, and sequencing, arose not from a single gene, but from a newly innate sense of logic. All human populations have language. All, no matter how small or remote, speak a language. About 6,000 languages have been identified on our planet. According to those linguists who study the matter, all are rule governed and all are of equal complexity. According to Noam Chomsky[176] there is a "deep grammar" that is universal to all language. The only controversial aspect to his ideas concerns the beginning of it. He discounts for language the usual evolutionary explanation for almost all aspects of human bodily or mental structure or physiology.

The Greek "miracle," as it has been called, also arose relatively suddenly and expanded rather quickly, and without any obvious or known precursors. I have long believed that it was the mixing of populations within less than a thousand years of this miracle that created new genetic combinations for on which natural selection might act. There is much support for such a view in the writings of widely respected evolutionary geneticists and biologists including O.E. Wilson and Charles J. Lumsden.

That, I believe, was the engine that drove the sudden advances of Greek culture. Chomsky offers no reason at all for the sudden appearance of rule governed language. But he, like most other sane, reasonable scholars would undoubtedly not entertain the possibility of extraterrestrials in our past. So I shan't mention it. Neither will I mention the slight suspicion I start to entertain that the same interlopers may have had a hand in the Greek miracle.

The biggest controversy over Chomsky, as mentioned above, is his claim of the sudden unexplained appearance of it. Other writers have gone to great lengths to show the higher probability of a gradual evolution. Few of the authors of such studies mention specific dates, but most, believing in a gradual beginning between one and two million years ago or longer do recognize a relatively sudden growth spurt. These spurts are usually subject to reasonable interpretation as beginning contemporaneously at, or shortly after, the beginning of the Upper Paleolithic. Though I use in this volume the approximate time of 50,000 BP, some of the linguists who do give time frames for this sudden growth, speak of 60,000 or even 200,000 years ago, the latter figure being obviously off target.

The substance of the growth, the spurt, is the development of syntax, a set of rules that governs the joining of words and sentences, and the sequence of the components in the sentences, and very importantly, an attribute known as recursion. It refers to the embodiment of one phrase or sentence in another, such as: "John saw a boat that he believed to be the one his friend sold." Previously, in earlier times, beginning perhaps millions of years ago, vocabulary of single words, "signs," began to emerge in human communications.

One of the more interesting accounts, or hypotheses, on the subject is found in a recent volume by Derek Bickerton.[177] He is a well recognized author and scholar in the field of linguistics. His own views have evolved over the years. His most recent, dating from 2009, is in his *Adam's Tongue*. He freely acknowledges the changes in his thinking, and his current views do indeed seem more deeply thought through.[178]

He compares the protolanguage before the great leap to modern language to be similar to pidgin. Pidgin, to use Bickerton's words, is "what people produce when they have to talk to other people but don't have a common language." You might use words of the other person's language that you happen to know, but you probably wouldn't put them together in the way you did in your own language. One reason would be that, being foreign words, you had to grope for them and then had them pop out one at a time in no particular order, and with big time gaps in between.

Another reason would be that you probably wouldn't know all the words you needed, even for the simplest sentence. That, says Bickerton is the closest we will probably ever come to knowing how it felt to be trying to communicate with another person at the dawn of language. But we will not here be further concerned with the dawn of language. It would seem that there must have been language used between humans from long before the Upper Paleolithic.

Our concern will be with the transition which we can assume was at least probably in the UP. Whenever it was it was when language became rule governed, when humans began to speak in a logical order according to rules. When it became recursive, whenever that was, it greatly expanded what could be said and conveyed, about things, people, thoughts, the present, the future, the past and countless other subjects and infinite nuances.

He claims however that our ancestors probably could have done much with words without grammatical structure.[179] What, he asks, do we gain by putting words into a modern syntactic form? The modern is longer and more verbose, the price we pay for eliminating obscurities and ambiguities and gaining a smoother flow of words. But which would be most important to thinking? He feels that given enough of the right words, our forebears might have done a little bit better than they did for a million years with the same old hand axe, as we have seen with Homo erectus (Chapter XIII) in the Lower Paleolithic.

Bickerton believes, in short, that protolanguage might not have reached even the *level of an "early-stage pidgin until our own species emerged." Further, he* conjectures that "the ability to connect words, to construct short and practical messages, emerged long before it became possible to link concepts into coherent trains of thought." The words had to come before concepts, not the other way around.[180]

Bickerton sees two ways in which words can be put together. One was like "beads on a string," the pidgin language. The other was by means of hierarchical, the way all languages are spoken today. When did this change occur? We don't know, he says, but his best guess is that it was at the earliest a couple of hundred thousand years ago. On what basis? Bickerton here betrays a disinterest, or lack of knowledge, in prehistory from the archaeological point of view. Steven Pinker, in his 1994 volume, *The Language Inst*inct had complained that archaeologists knew little about the work of psycholinguists (which he helpfully explained does not mean linguists who are psycho, but those who study psycholinguistics) but perhaps archaeologists could return the compliment. Bickerton mentions the "couple of hundred thousand years ago because,

> That's the earliest date so far suggested for the origin of our species. And it's around then that the first signs of really human behavior become manifest. Tools start to shape up a little, but it's not that. People are beginning to use ochre to decorate their

> bodies (or so we assume—they were using the stuff for something. That's for sure). Types of stone used for tool manufacture are found hundreds of miles from their sources, which suggests that some form of trade had started up. That meant contact between groups that probably didn't even speak the same protolanguage.[181]

His reasoning is sound, but his sense of time is a bit off. All of the things he mentions happened 50, or maybe 60, or maybe 45 thousand years ago, not one or two hundred thousand. That earlier time of Bickerton may mark the beginning of language in the bead stringing, or pidgin, era, but if syntax and hierarchy in language began with the changes he outlines, it would be better to ignore his suggestion of time and rely instead on the beginning of the UP, usually given as about 50,000 years ago. All of the changes summarized by Bickerton are the changes from the Middle to the Upper Paleolithic, spelled out in detail by Ofer Bar-Yosef in Chapter XIII above. Bar-Yosef and others place that transition at about 50,000 BP.

The objection to the suddenness of the syntax appearance in languages as described by Bickerton is well expressed by Eric Gans, of the Department of French at the University of California in Los Angeles. It is enlightening.[182]

> Bickerton's more recent *Language and Species* (1990) proposes, on the analogy of the distinction between ungrammatical pidgin and grammatical creole, both an early and a late origin for language. The early origin, at the time of *Homo habilis* [over 2 million years ago], would have involved the emergence of "symbolic reference," the linguistic sign, but not syntactic structure. Syntax, in Bickerton's view, could not have evolved gradually, since there are no examples of a language intermediate in syntactic complexity between pidgins, which he finds comparable to the utterances of young children as well as to those of

apes instructed in human language, and the natural languages of today. (It is a tenet of modern linguistics that all known languages, from those of the Australian Aborigines to contemporary English, are equally "advanced" and permit in principle of reciprocal translation.) Thus the emergence of syntactically mature language as we know it, which Bickerton situates at the time of late origin around 50,000 years ago, would have reflected evolutionary developments in the brain that were realized in language all at once in some inexplicable final mutation.

It is explicable, though probably not in a way satisfactory to most scientists. To continue with Gans:

Even if all modern languages derive from a common ancestor spoken around 50,000 years ago, there is no need to assume that this *Ursprache* [original speech] itself emerged in a single mutational leap beyond primitive pidgin-type languages.

Whether there is a need or not, the important issue is whether there are circumstances rendering sudden change likely.

We turn to a volume by Christine kenneally, also a scholar and author in the field of linguistics. Like most, if not all, linguists and archaeologists she states that Homo sapiens emerged in Africa about 200,000 years ago. According to her, archaeological and paleoanthropological evidence indicates that our species remained relatively stable for about a hundred thousand years. Between 60 and 80 thousand BP "there was a dramatic expansion of certain genetic lineages in the African population."[183]

At the same time there were striking changes in technology and culture, though there was no change physically. For one thing, she reports, there were many more types of unambiguous symbols. There were traces of art, including perforated shells used for jewelry. What she claims as the oldest known examples of human art were found in South

Africa's Blombos Cave. It consisted of two pieces of ocher with a hatching design carved into them. Shells had been clearly transported from other locations to sites where they have been found.[184]

She places the exodus from Africa by these humans, our direct fathers and mothers, at about 60,000 BP, and repeats the conclusion of many other scientists that "Everyone alive today descended from this small band of travelers." She states that these first modern humans may have traveled from North Africa and then split, some going north to Europe, at about 40,000 years ago, the others heading east to Asia. She notes, what has been uncontested, that by 40,000 BP Homo sapiens were sculpting from stone, painting in caves, and ritually burying their dead with grave goods.

How does language fit into this picture? Kenneally refers to Klein's position that these cultural changes, including modern languages all resulted from a sudden alteration in the human brain, resulting probably from a genetic mutation. This is a conclusion with which she does not completely agree. She states however that that "Even if you don't subscribe to the notion that a dramatically demarcated revolution occurred, it's clear that a major shift in the history of the human mind was taking place."

Keimeally devotes much space to the role of a gene known as FOXP2. Since its discovery in 2001, neurologists, linguists and other scientists have debated its role in language. Aside from recognizing that it does indeed play an important role, there is little agreement, even as to whether it affects vocabulary, syntax, or something else, or all of the above. Assuming for the time being, that symbolic speech (bead stringing) occurred at least 200,000 BP, and that use of syntax, as we use it today, occurred about 50,000 BP, we are most interested in knowing when this gene first appeared, or at least when it mutated to its current form. We will get little help here, but what little there is should be stated. It is part of the story.

We shared the original gene millions of years ago with mice and gorillas. There have, in humans, been two mutations of it according to Wolfgang Enard, who presented the history of the gene. With his colleagues he estimated that between 200,000 and 50,000 BP all humans had the gene. In the time frame of our genome, this is unusually rapid, which indicates "selection" which means it was highly adaptive. Kenneally states that the dates supplied by Enard coincide "perfectly with the acceleration of culture and the migration of modern Homo sapiens from Africa out across the world."[185]

It must be emphasized that there are at least two other matters that are the subjects of agreement by substantially all scientists: The first is that FOXP2 is only one of a number of genes, albeit an important one, affecting our ability to speak a language. The second matter of agreement is that all of the "hardware" necessary for human speech, the vocal organs, were in all probability, present in humans many thousands of years before 100,000 BP.

CHAPTER XVI

MUSIC

THEY PROBABLY SPOKE IN SENTENCES, but we do not, and cannot have proof of it. But they made music, and of that we do indeed have proof. It comes in the form of what are beyond doubt prehistoric musical instruments, mostly flutes.

Many decades ago a flute was excavated in Austria that was dated to 19,000 BP. A large number of those instruments, about twenty-two, were ultimately found in the French Pyrenees Mountains dating up to 30,000 BP. No archaeologist, or anyone else, was going to put his mouth on anything of that age, but exact replicas were made of some of them, from the same material, usually bird legs, or other bones of various aviaries, of the same dimensions and same number, placement and size of finger holes. They made music. We of course do not know what their music sounded like, but what was played on exact replicas of them by our modern flautists was indeed music.

It seemed inevitable that older flutes or other instruments would, sooner or later be found. It wasn't long. In September 2008 Nicholas Conard, an archaeologist from Tübingen University led a team that found it in a plot in one of the largest caves in southern Germany. It is the Hohle Fels (Hollow Rock) located in the Swabian Alps about 1750 feet above sea level, near the city of Ulm. It was there that Conard's group found twelve pieces of the wing bone of a griffon vulture. They put the puzzle together and found it was a flute. The bone was naturally hollow and was .3inch, or 8 mm wide. It was the eighth flute found in the area in recent times; four were from bird bones, four were from mammoth ivory. The one found in 2008 was 8.6 inches, or 22 centimeters, long, with five finger holes and a V shaped notched end.[186]

A flute fragment had also been found earlier at a nearby site named Geissenklösterle, and was dated to about 35,000 years ago. The mammoth-ivory flutes would have been especially challenging, the team said. Using only stone tools, the maker would have had to split a section of curved ivory along its natural grain. The two halves would then have been hollowed out, carved, and fitted together with an airtight seal.[187]

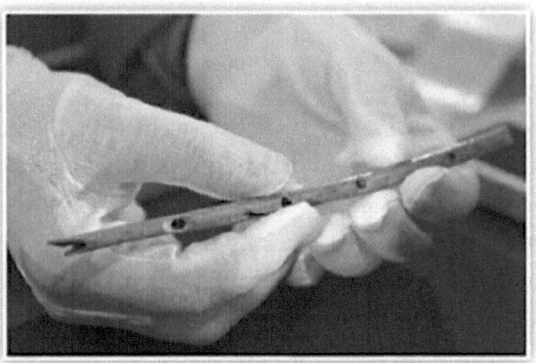

Illustration 39: Professor Nicholas Conard of the University in Tübingen shows a flute during a press conference in Tübingen, southern Germany. The thin bird-bone flute carved some 35,000 years ago and unearthed in a German cave is the oldest handcrafted musical instrument yet discovered.

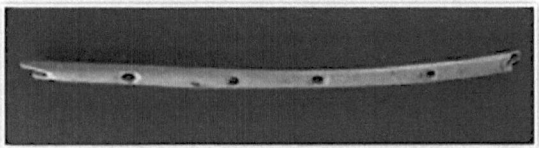

Illustration 40: Another view of the 35,000 year old flute made from a vulture's wingbone

The story of the Hohle Fels find was published in the journal Nature and the Associated Press, to whom Conard claimed it was "unambiguously the oldest instrument in the world." Other archaeologists agreed. A female figurine was discovered the previous May in the same layer of sediment in Hohle Fels. It was claimed to be the oldest known sculpture of the human form. Wil Roebrocks of Leiden University in the Netherlands, who was not part of the group discovering the flute, said that the two discoveries told us much about the people who were there at the time.

To ascertain that the instruments were dated correctly, samples were tested independently, using different methods at laboratories in England and Germany. Both facilities found the bone to be at least 35,000 years old.[188] To Roebrocks it suggested that modern humans had established an advanced culture in Europe 35,000 years ago. He claimed that the physical trappings of their lives, including musical instruments, personal decorations and figurative art, match the objects we associate with modern human behavior. "It shows that from the moment that modern humans enter Europe... it is as modern in terms of material culture as it can get."

April Noel, a Paleolithic archaeologist at the University of Victoria, Canada, and likewise not a participant in the discovery, studied the find and agreed that the flute was older than other previously discovered instruments. A functional morphologist, Jeffrey Laitman commented that the flutes show that the human society was becoming modern. They were not simply devoting their lives to finding food. The flutes "are telling us about intricate and delicate communication, bonding, social events that are going on."

Conard told *National Geographic News* that early human budding culture might have helped them survive, and that the ancient flutes are evidence for an early musical tradition that likely helped them communicate and form tighter social bonds. Hence music, he said, may have been important in maintaining and strengthening Stone Age social networks, allowing for greater societal organization and strategizing.

His team argued that the emergence of art and culture that early might explain why early modern humans survived, and why Neanderthals, with whom they co-existed at the time, became extinct. Chris Stringer of the American Natural History Museum told BBC World Service in the same vein:

> These flutes provide yet more evidence of the sophistication of the people that lived at that time and the probable behavioral and cognitive gulf between them and Neanderthals. I think the occurrence of these flutes and animal and human figurines about 40,000 years ago implies that the traditions that produced them must go back even further in the evolutionary history of modern humans—perhaps even into Africa more than 50,000 years ago... But that evidence has still to be discovered.[189]

Neanderthals also occupied much of Europe as the modern humans moved in from Africa. The possibility that the instruments were those of Neanderthals was rejected by all who considered the matter. The absence of any musical instruments or almost any other trappings of artistic creative enterprise over a period of hundreds of thousands of years was among the many reasons. The creativity of modern humans, on the contrary, continued long after the Neanderthals became extinct, something that occurred about 10,000 year after the modern humans arrived in Europe.

The bird-bone flute probably produced a range of harmonic tones similar to the modern flute, according to a specialist in ancient music, who reproduced a Stone Age flute made of ivory to see how the original might have sounded.

CHAPTER XVII

ONE FACTOR IN TRANSITIONS

THE SUBJECT OF MIXING OF genes of populations who had previously lived completely separate from each other, and the subject of epigenetics are both a vitally necessary part of the picture here. So perhaps it would not be amiss to briefly summarize both subjects: 1) the evolutionary results of mixing of genes of long separated populations (or those who have never lived together), and 2) subject of epigenesis and its role in spreading new traits rapidly through a population. This was first discussed in Chapter VI.

According to Milford Wolpoff, gene flow through the mixing of long separated populations acts a "creative force in the evolutionary process."[190] Such mixing, says G.A. Harrison, "restores vigor to isolated groups,"[191] Lumsden and Edward O. Wilson speak of a "Theater of Opportunity" upon which natural selection cans act.

What such scholars claim is that the mixing of genes of groups long separated, and who have developed their own respective genetic signatures, creates new genetic combinations, those conveying advantageous change to the organism being most likely to survive. Genetic change however does not act on an entire population at once.

Lumsden and Wilson have addressed the issue of how new behavioral or cognitive traits can spread among an entire population, and the time required for it.[192] Playing a key role in recognizing the rapidity by which new genetically based traits, especially cognitive traits, can spread through a population, is a relatively new understanding of "epigenetics." They furnish a time frame during which genetic changes can be expected from it. They claim that with sufficient evolutionary pressures genetic change can proceed as rapidly as cultural change and can result in partial replacement by one form of a gene, that is, an allele, by another within ten generations or about two to three hundred years. As a more general matter they believe that a fifty generation. or a "thousand year rule" suffices.

Contrary to the belief still held by many biologists, they believe that evolution of cognitive traits has continued through modern times.[193] They thus speak of "co-evolution," meaning the interactions in which biological factors generate and shape culture and in which biological traits at the same time are altered in response to cultural innovation.[194] This latter proposition has long been anathema to most biologists, the germ line being allegedly impervious to change in the genes of the soma, that is, the body except for the reproductive cells.

As Nessa Carey puts it:

> In the 21st Century it is the new scientific discipline of epigenetics that is unraveling so much as what we took as dogma and rebuilding it in an infinitely more varied and more complex and even more beautiful fashion... The world of epigenetics is a fascinating one. It is filled with remarkable subtlety and complexity.[195]

Well said, and obviously true enough. But the examples deal almost exclusively with minor behavioral or physical traits. There are no examples of either phenomena resulting in such momentous change as the great transition that is our subject. These subjects are offered here as something to be kept in mind for whatever probative value one might feel they, or either of them, may have on the changes described in the next few chapters.

CHAPTER XVIII

THE TRANSITION IN THE FERTILE CRESCENT

It happened in a hill in southeastern Turkey, south of forested mountains, west of the plain of Harran, and north of the Syrian border visible 20 miles away. It is known, in Turkish, as Göbekli Tepe, in English, as Potbelly Hill. The border points to the ancient lands of Mesopotamia and the Fertile Crescent. According to archaeologist Klaus Schmidt, that hill is the spot where all civilization started. Even those most awed by the brilliance of the Upper Paleolithic in Europe do not call their culture a civilization. By our definitions, people who live in caves, lacking in monumental structures, cannot be termed civilized. So there is sound basis for Schmidt's claim. On what evidence does he base it?

His evidence is a large and stately temple complex, which, he says, may be the first thing humans ever built. It has been dated at 11,500 years ago, 4500 years before the Great Pyramid; 5500 years before the beginning of Stonehenge, at least according to the commonly accepted dates. The questioning of those dates by advocates of earlier ET presence on Earth, right or wrong, has no bearing on the significance of this structure.

According to Schmidt it predates villages, domestication of animals, and even agriculture. Schmidt believes it was built by hunter gatherers at the end of the ice age, and was the beginning of farming and village life.

He spent more than twelve years of painstaking work here and is convinced that what he uncovered is proof that it is a huge ceremonial site, one where hunter-gatherers built a complex religious community. He has found carved and circular polished stones, the largest being 30 yards across, and featuring T shaped pillars up to 17 feet in height. According to carbon dating they are among the oldest monumental artworks. Elaborate carving can be seen on about half of the 50 pillars. A few are abstract symbols, but the site contains mostly graceful, naturalistic sculptures and bas-reliefs of the animals that were central to the imagination of hunter-gatherers. There are wild boar and cattle together with lions, foxes, and leopards, totems of power and intelligence. Many of the biggest pillars are carved with arms, including shoulders, elbows, and jointed fingers.

Among his other finds are terrazzo flooring and double benches. In 2009 he found the third and fourth examples of the temples, and ground-penetrating radar indicates 15 to 20 more such ruins. Ian Hodder, director of Stanford's archeology program has spent decades on rival Neolithic sites. He enthusiastically called the finds at Potbelly hill "fantastic, highly refined art… Many people think that it changes everything… It overturns the whole applecart. All our theories are wrong." It could indeed turn some theories upside down, even more than Hodder was thinking of. It was the need to build and maintain this temple, claims Schmidt, that drove the builders to seek stable foods like grains and animals that could be domesticated, and then to settle down to guard their new way of life. He claims that "The Temple begat the city," the reverse of earlier theories.

And so it was. The accepted chronology described a "Neolithic revolution" about 10, to 12,00 years ago in which farmers appeared first, then created villages, cities, specialized work, government, writing, and ultimately, organized religion. Thinkers and philosophers argued that cities came first, and only thereafter the high religions with their great temples, something still taught in elementary schools. But if Schmidt is correct, religion may be less a product of culture than a cause of it; less a revelation than a genetic inheritance. According to Patrick Symmes, "The archeologist Jacques Chauvin once posited that 'the beginning of the gods was the beginning of agriculture,' "and Göbekli may prove his case.

Schmidt says that the religious purpose of the site is implicit in its size and location. "You don't move 10-ton stones for no reason... Temples like to be on high sites... Sanctuaries like to be away from the mundane world." Contrary to most discoveries from the ancient world, Göbekli Tepe, was found intact, the stones upright, the order and artistry of the work accessible even to the un-trained.

We come to aspects of the discovery that perhaps bear most relevance to out subject. We quote Symmes: "The T shapes appear to be towering humanoids but have no faces, hinting at the worship of ancestors or humanlike deities. Symmes also quotes Johns Hopkins archeologist Glenn Schwartz: "In the Bible it talks about how God created man in his image." These "humanoids" were created, it must be noted, perhaps 8,000 years before he Bible was written. Göbekli Tepe, continues Schwartz, "is the first time you can see humans with that idea, that they resemble gods." Rarely, even to the present day, when God, or gods are pictured at all, do they appear in human form.

Schmidt sees this temple as the end of a 140,000 year reign of the hunter-gatherers. He pointed out the evidence of that transition. "The people here invented agriculture. They were the inventors of cultivated plants, of domestic architecture," he says. Genetic mapping, he emphasizes, shows that the first domestication of wheat was in this immediate area—a few centuries after Göbekli's founding. Animal husbandly also began near here—the first domesticated pigs came from the surrounding area in about 8000 B.C., and cattle were domesticated in Turkey before 6500 B.C. Pottery followed. Those discoveries then flowed out to places like Catalhöyük. the oldest-known Neolithic village, which is 300 miles to the west.

One of the sources of Schmidt's claim for the religious significance of the site is the absence of traces of daily life. There are "No fire pits. No trash heaps. There is no water here." Everything from food to flint had to be imported, so the site "was not a village," Schmidt says. Since the temples predate any known settlement anywhere, Schmidt concludes that man's first house was a house of worship: "First the temple, then the city,"

Says Symmes: The real reason the ruins at Göbekli remain almost unknown, not yet incorporated in textbooks, is that the evidence is too strong, not too weak. "The problem with this discovery," as Schwartz of Johns Hopkins puts it, "is that it is unique." No other monumental sites from the era have been found. Before Göbekli, humans drew stick figures on cave walls (and, it must be added, they created splendid paintings, engravings, and sculptures deep within the recesses of European caves, though no structures, and no human forms) shaped clay into tiny dolls, and perhaps piled up small stones for shelter or worship. Even after Göbekli, there is little evidence of sophisticated building. Dating of ancient sites is highly contested, but Catalhöyük is probably about 1,500 years younger than Göbekli, and features no carvings or grand constructions. The walls of Jericho, thought until now to be the oldest monumental construction by man, were probably started more than a thousand years after Göbekli. Huge temples did emerge again—but the next unambiguous example dates from 5,000 years later, in southern Iraq.

Perhaps most significant of all was one of the things that so entranced Glenn Schwartz: gods that looked like humans, as do the Nordic aliens, we might add, from somewhere in the cosmos.

CHAPTER XIX

TRANSITIONS IN GREECE AND THE RENAISSANCE

Two of the most significant and, relatively speaking, dramatically sudden episodes of change in human thought, feelings, and behavior, occurred in historical times, millennia later than he two we have seen this far. One was that occurring in ancient Greece beginning in the fifth Century BC, and continuing in ancient Rome. The other was the European Renaissance beginning in the 15th Century. Each of those creative episodes involved quantum leaps in attributes that have been collectively called humanism. Each seemed to grow out of nowhere, exhibiting sudden changes in the cultures out of which they grew, and were largely in opposition to attributes of other contemporary cultures. From these two historical episodes we have written records and a plethora of artifacts, hence less need for speculation than does the transition to the Upper Paleolithic, or even that from the Fertile Crescent.

Those two episodes in historical times had much in common. Each of them were marked by tremendous increases in literary style to include, metaphoric and symbolic language; individuality and expressiveness in artistic depiction of the human face; philosophy involving analysis and speculation about the human place in the world and reasoned rules of ethical behavior; recognition of individuality and respect for the worth of the individual; and novelty and innovation. There was also, in ancient Greece the beginning of three dimensional perspective in art with the advent of foreshortening, namely 3-d perspective as applied to a single object. Later, in the Renaissance, came perfection of 3-d perspective of entire visual scenes, resulting in a newly discovered beauty in the view of nature for its own sake. How might these two remarkable episodes have come about?

There have been many scholarly attempts to explain them. None seem to be able to withstand even the most cursory analysis. There is little need to go into any detail in either case. The explanations seem to march in lockstep with the unfortunate and misguided specter of political correctness. A more logical and undoubtedly more correct analysis comes from evolutionary biologists whose explanations we will examine shortly. We note first the complex mixture of populations that preceded each of these creative episodes, and the possibility, perhaps the likelihood, that the mixtures had causal connections to the remarkable spurts of progress in humanism.

The earlier episode, that of ancient Greece, has most often been attributed to its "fortunate" location and the trade with other cultures it facilitated. Such a view ignores the same facility and amount of trade engaged in by the Phoenicians, among others, not to mention that the great Greek accomplishments in literature, art and philosophy, and in human relationships, bear no evidence of influence from any cultures in its past or contemporary with it. Such a view ignores also the opportunities that others, such as the Turks, had for far ranging commerce, but, after 500 years, still did not avail themselves of the opportunities.[196] Just north of Greece, the Balkan Peninsula had rich deposits of marble and clay, but produced no original art. As the Greeks themselves recognized, they did not learn from others; others learned from them.

The later episode, the European Renaissance is most often attributed to the rise of the business class, with no thought given to the clear possibility that such development might have been a result of the change in cultural development rather than a cause of it. How the rise of a business class could have resulted in the works of the great artists, writers and philosophers of the Renaissance would be difficult to explain.

We look first at the Greeks. It was probably sometime before 1550 BC that Greek speaking Indo-Europeans from the Hungarian plains moved into the southern part of the Balkan Peninsula and along the western coasts of what is today the nation of Turkey. The fusion was completed probably within a few generations, but it was the Greek language that became dominant as it did wherever the Greeks settled.[197]

Among the invading tribes were the Ionians, probably among the most important forebears of the later Athenians. They occupied wide areas including parts of the Peloponnesus and to the Mediterranean Islands known as the Cyclades and the Aeolian. By 1550 B.C. there arose a remarkable culture at the city of Mycenae. Among other important discoveries, it is noted for the oldest evidence of wheeled chariots in Greece, mighty structures of huge stone walls, many-sided masonry, and gold and fine jewelry, probably from plunder. The full glory of the city lasted for about 200 years.

Also pushing south into Greece, by about 1200 B.C., the approximate beginning of the Iron Age, were the Dorians. Homer and others called them Achaeans. They assimilated into the Greek populations where they settled, including the Peloponnesus, with its people of wealth and luxury, and Thessaly, much less so, in the north. For over a hundred years, there were migrations and displacements of Greek peoples and others who, though not originally Greek, were assimilated through intermarriage. It is a truism among anthropologists that when two populations have the opportunity to interbreed, they do.

It has been noted that widespread intermarriage took place, and that the invaders wedded the daughters of native families, an intermingling that according to historian Jean Hatzfeld,[198] was instrumental in the rise of Greek civilization. It has also been called by others[199] a vitalizing factor in the creative energy of the later Greeks, whose civilization was credited by Germain Bazin[200] to the ferment of these successive migrations.

The southern movement of the Dorians affected a group, known as Illyrians who were driven south from the Balkan Peninsula, and the Thracians, who were forced east into Asia Minor. Both were Indo-Europeans, but not Greek. Still other populations that were Greek were pushed south into the Peloponnesus and the Sporades islands. In the meantime, from about 1400 B.C. there were attacks on the Peloponnesus by populations from Mediterranean Islands, known as "people of the Sea."

Indo-European, Achaean, Dorian, Mycenaean, Ionian, Thracian, Illyrian. It may mean very little, or may mean nothing, but, the descriptions of the Nordic aliens considered, it may be worth mentioning something of the physical appearances of the groups that formed the Greek population beginning about 600 B.C. We make special mention of the hair, skin and eye coloring for reasons that will be clearer later, with proviso that no firm conclusions could be drawn from these observations. They are offered for whatever probative value they may have, if any.

The Dorians are often described as fair haired. Historian J.B. Bury claims that the description was a mistake that started with Homer. He claims also that among the later Greeks, two marked colors existed, namely light and dark. The light complexioned, he says, being rarer and more admired.[201]

The Thracians were described by some contemporaries as a ruddy, blue eyed people. The philosopher Xenophanes wrote in the 6th Century B.C. that the gods of the Thracians had blue eyes and red hair "like themselves." In the *Iliad* of Homer (Book V, line 500), Demeter, the goddess of the fruits of the field, is described as "fair-haired." Often in the Odyssey he writes of the goddess Athena as grey-eyed, but she is described in the *Larousse Mythology* (p. 107) as the goddess of the "brilliant eyes). Homer (Book i, line 102) describes Meneláos, King of Sparta, as the "red-haired king." Sappho has described Helen of Troy as golden-haired.

It is interesting that nowhere in Homer, or in many other ancient descriptions of physical attributes, is there mention of dark haired or dark eyed coloring. It may be a clue that those were attributes of the large majority. If so, we are entitled to wonder where those blond, or red, hairs and blue eyes may have come from. History does not record much about Scandinavian influence, or of interbreeding with them.

Of further interest is a sculpted head found in the Old Palaestra on the Island of Delos in 1912. It is believed to date from the early 1st Century B.C. Though many Greek sculptures from Greek antiquity are missing eyes and coloring, this one appears to be an exception. It contains clear evidence of blond hair and brilliantly blue eyes.[202]

But if blond hair was the exception in ancient Greece, in that land's folklore or mythology it is not. Among the blond haired goddesses are Athena, Hera, Aphrodite, Peleus. Achilles, Meleager (said to be second only to Hercules), Agamede (practitioner of witchcraft who knew the healing power of all plants), and Rhadamanthys[203], son of Zeus and Europa. We will see, and may wonder about the reason, that gods seem so often to be portrayed as blonds, even when the population may have relatively few. Perhaps it is not all based on imagination.

But maybe one of the more interesting, and more significant bits of information comes from this phrase from Homer's Iliad (Book iv, line 50), wherein he refers to Hera, sometimes called the "white armed goddess," as the "ox-eyed lady." The ox's pupil is always brown, and it is doubtful he is referring to color here. What is different about the ox's eyes is not the color but the fact that the pupil is oval rather than round as in the case with humans. Oxen are not the only animals with the oval iris. Cats, for one, also often have the oval shape. The oval, where it exists, is vertical, and there have been a number of descriptions of Nordic, and other ETs as having vertical ovals. They are also often referred to as slits.

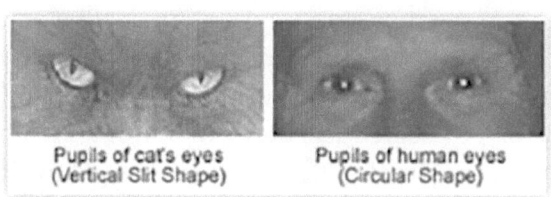

Pupils of cat's eyes (Vertical Slit Shape)   Pupils of human eyes (Circular Shape)

Illustrations 41: Although some animals' eyes are basically structured in a similar way to human eyes, they may appear to be very different. E.g. differently shaped pupils of cats compared with people.[204]

Grey extraterrestrials have elsewhere been described as erect standing, biped, small, thin build, large and rounded rear skull area, head larger than humans with inverted triangular shape, absence of auditory lobes and nose, absence of body hair, large oval tear-shaped eyes which are opaque black with vertical slit pupils.

Let us turn to the later period of quantum leap in human thought, feeling and emotion, known as the European Renaissance. We looked back from the Greek fluorescence about a thousand years, in accordance with the calculations of biologists that it required at the most such a time frame in which new alleles could become dominant in a human population. Let us look back the same distance into the past from the rise of the Renaissance in the 15th Century.

That brings us to the incursions of the tribes, often referred to as barbarians, of northern and eastern Europe, into the area of the Roman Empire. The Romans, in many ways, had equaled or surpassed the Greek humanism, equaled it in literature, surpassed it in 3-d perspective in art. This is a case of acculturation, which has happened between almost all cultures in more recent times. When we do, we are met with a more numerous and bewildering assortment of populations who had been living separate from the people of the empire for thousands of years, and more than likely, from each other. There were the Cimbri, the Cherusci. Ostrogoths, Visigoths, Turanians, Huns, Vandals, Burgundians, Franks, Lombards, Bavarians, Swabians, Saxons, Gepidae, Bulgarians, Rugi, Breoni, and perhaps scores of others.

Most of them did not come to conquer, plunder, and leave; most came to stay and settle in. There was no written history of any consequence during the thousand years before the 6th Century B.C. and what we know of Greek history before that time was pieced together by archeologists and anthropologists. But by the time of the Romans, there was written history to spare. The story of Europe after Rome's fall is one of destruction, slaughter, perfidy, and general chaos. Out of it came order, not to mention stagnation, supervised by the Christian Church. It was a thousand years of rigid rules of behavior and thought, but also of combat between armies of tyrants.

About 800 to a thousand years later, there was stirring of something obviously new, love poetry, or *amour courtois*, and by the 15th Century there blossomed a rebirth of all that was best in Ancient Greece. So what about the transition to the fully modern humans of 50,000 years ago? Most assuredly there are no written records. Nonetheless, could this remarkable transition in Africa have been preceded by a mixture of long separated human populations such as those that preceded the Greek fluorescence and the European Renaissance? It hardly seems possible.

The entire source population where the modern behavior first took hold was a small group, originally in northeast Africa, estimated at only 5000 people. About those times we can hardly say anything for certain. But there is little if any reason to believe that there was any great migration event between them and any outside group coming suddenly after a long period of separation. Their world was not like that. They did not live in settled communities, but in small bands of hunter-gatherers, and their lives were dominated by continuing warfare, not the kind conducive to, or even permissive of interbreeding.[205]

Though such a subject is never discussed in the scholarly literature, we can assume that the mixture of populations that have no common source, never "separated" because they never lived together. But they would have even more potential for genetic change, for new combinations of genes, and a greater "theater of opportunity." How could something like that happen? Could the outside race be the Nordics, the tall whites, who have been reported still to be among us? Let us look, cautiously, into that possibility.

PART THREE

THE NORDICS

CHAPTER XX

DO THEY TIE THE THREADS TOGETHER?

There are many different descriptions of extraterrestrials, especially in recent times, too many to list, let alone describe. Suffice it to say, according to all descriptions, most ETs look very little like us, and most of the remainder look nothing like us at all.

There have been many reports of hideously strange looking creatures, obviously not human, enough that they could, if accepted at face value of the reports, comprise twenty or more races. It would be worthwhile, before looking at the Nordics, to look at a few of the others for comparison. We will look only at those few whose presence here has been attested to by witnesses of proven, or apparent, high credibility, or by a sufficient number of witnesses whom we would have no reason to doubt.

The Greys, in recent times, have been by far the most numerous. They are the ones whose short, thin frames and large bulging black eyes have become in popular usage the icon for all ETs. The sheer numbers of reports of these entities should carry some weight, but we should hear also from some of the more scientifically minded. With the Greys we must keep in mind that there exists the high probability that some of them, though resembling EBEs (extraterrestrial biological entities) are not biological entities at all. Many are probably robots. As in all other fields of technology this race is far ahead of us in robotics, and just as our technicians aim to make robots that look like us, the Greys may well have done the same. It may be that even trained doctors and scientists, looking at creatures from another world, could not tell the difference, uninformed as they are in either the physiology of the biologic ones, and of the structure of the robots, if they are indeed robots. Without doubt, some of them are, as they lack the internal organs that would seem necessary on any planet

Bearing that in mind, we start with investigations following the crash of a UFO near Roswell, New Mexico in early July 1947. One description, also second hand, comes from an unnamed doctor who participated in the autopsy of one of the bodies and related the findings to Dr. Len Stringfield. A second physician saw another body, which he and others also described to Dr. Stringfield. From these sources Dr. Stringfield later wrote a description of the entire body, which was reported and summarized by Thomas J. Carey and Donald R. Schmitt in *Witness to Roswell*.[206] "I learned of its internal chemistry and some of its organs-or by human equation, the lack of them," says Stringfield. It is with the findings concerning the internal organs and their structure that we are most concerned, but we begin with a more superficial description, namely that of the outer appearance. Carey and Schmitt reported that according to Stringfield,

> The being was humanoid, 3 to 4 feet tall, and weighed 40 pounds. Its head was proportionally larger than a human head. It had two large round eyes. Its nose was vague with only a slight protuberance. Its mouth was a small slit and opened only into a slight cavity. The mouth apparently did not "function as a means of communication or as an orifice for food ingestion," and there were no teeth. There were no earlobes or "protrusive flesh extending beyond apertures on each side of the head.

PALE SKIN, GIANTS, AND THE GREAT TRANSITION

The body had no hair on its head, though one of Stringfield's medical sources said it was covered with slight fuzz. The neck, torso and arms were thin and the hands reached close to the knees. A slight webbing effect between fingers was noted by three observers. Skin color was described variously as beige, tan, brown, or pinkish gray. Texture was described as scaly or reptilian and as stretchable, elastic, or mobile over smooth muscle. Under magnification, says Stringfield, "I was told the tissue structure appears mesh-like," suggesting to him "the texture of granular-skinned lizards, such as the iguana and chameLeón."

Among others describing the reptile-like type was one rather distinguished observer named Werner von Braun, the former German rocket scientist, at the time in question, working for the U.S. government. Shortly after the Roswell crash, he was taken to the site from his employment at the White Sands Proving Ground, New Mexico. Included among his observations of the bodies of the occupants, he said that their skin was grayish and reptilian in texture... similar to the skin texture of rattlesnakes he had seen several times at White Sands.[207]

Were they Greys, or a separate group called by some the Reptilians?

Some very interesting statistics have been reported. Between 1982 and 1995 people in another part of the country, the Hudson Valley, were reporting a plethora of sightings of UFOs. During the late 1980s and early 90s there was, in fact an acceleration of reports of 'close encounters,' usually referring to anything from sightings of UFOs to sightings of 'animate beings'. The total number of sightings recorded during that period was 7,046. They were broken down for statistical purposes into various categories, including the category of our interest, the sighting of humanoids. A total of 256 such sightings were reported. For the entire reporting period examined 48.4% were described as short grays; 9.8% as tall grays; and 40.2% as "reptile; 1.6% were called human, probably the Nordics, or Tall Whites, if the two are not in fact one. That is a very high percentage of Reptiles, and a mystery, considering how seldom reptile-like skin has been reported in the UFO literature.

There are a few other sightings that are obviously neither Nordics, Tall Whites, nor Greys, but have sufficient supporting evidence to warrant inclusion as typical of the differences between them and the Nordics, and the obvious differences from natives of our planet. I will mention two.

There is one case from Kentucky in 1955. It involved the Sutton family, consisting of eight adults and three children.[208] About 7:P.M. one of the children went outside to get water out of the well, but soon dashed back into the house to report an object falling from the sky. The family dismissed the report until the dogs started barking. Two adults went out to investigate and were shocked to see a glowing creature about 3 ½ feet tall. Unlike the usual case with the Greys, this creature did not retreat, but approached them. The family barricaded themselves in the house. The subsequent events include seeing one of the uninvited at the window, and another on the roof. Shot guns were fired at them, probably hitting one but without much effect except to cause the strange pair to retreat. The family, piling into two vehicles drove post haste to the local police.

To the police they described the invaders as having a huge head, wide set eyes, and elephantine ears. This last item is also contrary to most all descriptions we have thus far seen, especially of the Greys. He also had abnormally long arms ending in clawed hands that hung at this side. He walked with a stiff gait, which the Suttons later determined resulted from the fact that his legs did not bend at the knees. Further, he did not seem to have a neck, and Mrs. Sutton said he appeared to be dressed all over in tin foil with the shiny side out. Not at all our typical Earthling.

We look next at an episode in Caracas, Venezuela, beginning at about 2:A.M on November 28th 1954.[209] Gustavo Gonzales was a panel truck driver, leaving Caracas with his helper Jose Ponce. They soon found their way blocked by a glowing object, shaped like a disk, about ten feet in diameter, hovering about six feet above the street. They stopped and stared, dumbfounded, then left the cab and walked toward the object. When about 25 feet from it they were approached by a small occupant, about 3 feet tall, hairy, animal like, and his eyes were fierce as they glowed yellow in the track's headlights.

Gonzales grabbed the mysterious looking fellow, and lifted it off the ground. He later estimated its weight at about 35 pounds. But with a show of great strength, the little creature twisted himself loose, and in the process, gave Gonzales a shove that sent him sprawling across the street. Ponce, showing somewhat better judgment than his boss, ran to the police station, about two blocks away.

Meanwhile Gonzales, up on one knee, got his knife out as his undersized adversary returned to the attack. Gonzales could now see that in lieu of hands his opponent had webbed extremities with about inch long claws. With these he raked Gonzales, even as Gonzales was trying to drive the knife into the other's shoulder. According to Gonzales, the blade glanced off as though it had struck steel. Then a second hairy creature emerged from the craft and, with a beam of light, blinded Gonzales, who now thought his life was over.

But only the fight was over. In a few moments he regained his vision, in time to see the craft rise above the trees and disappear from view. Gonzalez now also went to the police station. He was torn, bleeding and thoroughly terrified. The first reaction of the police to this story was disbelief. No surprise there really. But the doctor that the police summoned found that both men were indeed in a state of shock, and that neither of them had been drinking. Gonzales was treated for a long deep scratch on his left side and was given a sedative.

There was an independent eyewitness. A doctor on night calls had also seen the UFO and submitted a corroborating report to the police with the understanding that his name be kept confidential. No reason for surprise there either.

Assuming the tale is true, and I see no reason to doubt it, what do we make of it? These two occupants were not the only sightings of hairy creatures from UFOs, and some will be described as fairly tall. Justified though they obviously were, these two also seem more aggressive than others we have seen. Are they related to the ones we just saw in Kentucky, scaring the Suttons out of their wits? Are they from different planets? Different races from the same planet? No answers. Very frustrating. Almost all of these groups, many of which we have not described at all, are obviously visiting from somewhere else in the cosmos.

All, in fact, but one. or possibly two. Though by some, the Nordics and tall Whites are treated as separate races, I put those Ufologists in the category of "splitters." I will play the role of the "lumper" as I feel from the descriptions there are more similarities between them than differences.

The Nordics are a group that has been described as very human looking, mainly, at least according to their coloring, they look like people of Nordic ancestry. But in another and more important sense, they look like all of us, namely they have, for starters, two legs, two eyes, two arms, a nose and an outer ear. Apparently, on close examination they could not easily be mistaken for ordinary folks. Their features, at least on close examination, are said to be much sharper than ours. Their height is anywhere from 5'6" to 9 feet tall, those of the high end, being real attention getters. Yet some, presumably in the shorter range, have mixed in with humans without attracting attention. The Nordics have pale skin, described sometimes as bluish white. The eyes have sometimes been described as having pupils that are ellipses, vertically oriented making them look like slits, but are at other times, described as circles like ours.

Though seen in the United States far less often than the Grays, they have been seen often enough, world wide, by witnesses with enough credibility, that only those who reject any idea of extraterrestrials at all, can ignore such reports. We need to look at them.

CHAPTER XXI

NORDICS: MCCLELLAND AND OTHERS

Let us start with one example. A Former employee of NASA, Clark McClelland, retired from that space agency in 1992 after 34 years of service. He had worked on the Mercury, Apollo, Skylab, space shuttle, and the International Space Station as a space craft operator (ScO). Upon leaving NASA's employ in 1991, he received some very commendatory letters from a number of persons, well known in the space industry in various capacities.

There was, for one, Walter Cronkite, who gave him a photograph inscribed, "To a space pioneer and a top Space Craft Operator. With admiration for your work"; A letter from Dr. Werner von Braun, space flight pioneer and NASA scientist said "Your gift of knowledge concerning space exploration, astronomy, and cosmic life is most impressive. I've always enjoyed our talks… Clark, it is as impossible to confirm them (UFOs) in the present as it will be to deny them in the future"; Brian Duffy, NASA astronaut and space shuttle mission commander wrote "Thanks for the great orbiter (Shuttle) you gave us. It ran better than my lawn mower!"; One from Dr. Sally Ride, NASA astronaut and first

American woman in space said "Thanks for letting us borrow YOUR [emphasis in the original] orbiter"; Kathy Thornton, NASA astronaut and space walker who helped repair the Hubble Space Telescope in orbit wrote "Clark, you were a GREAT [emphasis in the original] Space Shuttle ScO."

From these testimonials we might anticipate that McClelland would receive honors and generous retirement income. He received neither. We might also think that his records of service in NASA would be very available. They apparently were not. A number of researchers have tried, with either very limited or no success. Why?

In 1991, a year before leaving NASA, he witnessed something very startling. During a space shuttle mission in 1991, he states he witnessed an extraterrestrial communicating with tethered astronauts:

He described it in an affidavit he later executed:

> I, Clark C. McClelland, former ScO, (space craft operator) Space Shuttle Fleet, personally observed an 8 to 9 foot tall ET on [my]27 inch video monitors while on duty in the Kennedy Space Center, Launch Control Center (LCC). The ET was standing upright in the Space Shuttle payload bay having a discussion with two tethered US NASA Astronauts! I also observed on my monitors, the spacecraft of the ET as it was in a stabilized, safe orbit to the rear of the Space Shuttle main engine pods. I observed this incident for about one minute and seven seconds. Plenty of time to memorize all that I was observing. It was an ET and Alien Star Ship! A friend of mine later contacted me and said that this person had also observed an 8 to 9 foot tall ET inside the Space Shuttle crew compartment! Yes, inside our Shuttle! Both missions were DoD (Department of Defense) top secret (TS) encounters![210]

PALE SKIN, GIANTS, AND THE GREAT TRANSITION

Illustration 42: Recreation of scene as observed by Clark McClelland on TV monitor in 1991 at the Kennedy Space Center, Launch Control Center.

Did NASA ever deny the presence of the ET on the space shuttle? No. Instead McClelland was severely punished for it. He was denied the pension he had earned for his many years of service. He was "blackballed" by NASA and unable to get work in the aerospace industry. He finally took a job involving menial work at a low salary. Though he claims that another NASA spacecraft operator told him he also witnessed the tall creature tethered on the space shuttle, he will not reveal the other's name, lest he subject him to the same treatment. Obviously, this vindictive action was not meant as punishment for a wild-eyed rant or a hallucination. It was something that could be easily denied were it not true. What NASA did was obviously meant as a warning to other potential whistle-blowers, and strengthens the belief that it was probably all too true.

The height corresponds with that often reported for Nordic E.Ts. What about the rest of the physical features? In a later interview with an Australian journalist, McClelland added, "It was an ET and Alien Star Ship. It was a tall creature, about eight to nine feet tall. It had a humanoid body shape with two arms, two hands, two legs, two feet, a slim torso, and a normal size head for its size. The color of its skin I could not determine. It appeared to have two what appeared to be eyes, but it was not detailed enough for any other comments."

He also said that he had no idea how it communicated, but that it moved his arms a lot.[211] Except for the height, it sounds a lot like us.

In July 2008, McClelland posted further information on the World Wide Web,[212] claiming that his observation was on his 27 inch video monitors while on duty with the Kennedy Space Center Launch Control Center (LCC). He states that the ET was standing upright in the Space Shuttle Payload Bay speaking with two tethered US NASA astronauts. He also observed the spacecraft of the ET, which was in a stabilized, safe orbit to the rear of the engine pods. He added further to his original affidavit that the ET moved its arms a lot, almost like giving instructions. "The helmet was not as large as our two NASA astronauts, and had a viewport to look forward. It had a small, perhaps a communication device, attached only to the right side of the helmet. I saw no oxygen tank(s). It [the ET] had a wide belt wrapped around it. It did not appear to be tethered as the two astronauts were to the sides of the shuttle structure. I observed nothing that appeared to be a weapon." He also added that he could not see the skin of the ET.

George LoBuono, writing in the journal *Exopolitics.* in an article entitled "Determining Human Relations with Aliens,"[213] quotes career Airman Charles Hall, author of four volumes on the subject of ETs at Nellis Air Force range, to the effect that he, Hall, witnessed how Air Force brass condoned the presence of "Tall White" aliens who were given a small base and supplies in the southeast corner of the Nellis Air Force range starting in 1954, or perhaps earlier.[16] Steven Wilson, about whom we will hear more shortly, Clifford Stone, and Michael Wolf all stated that they saw aliens working inside US bases during their military careers.

In addition, according to LoBuono, Army Sgt. Robert Dean claims he saw a secret NATO document to the effect that senior political and military leaders had mixed with human-looking aliens who could blend into a human crowd, unnoticed. Dean said the NATO command feared such aliens could infiltrate high public positions without being detected, a suspicion that Col. Corso said haunted the Pentagon. More recently, a computer hacker Gary McKinnon claimed to have viewed internal US military documents about "non-terrestrial officers" and "fleet-to-fleet transfers." The US government prosecuted McKinnon for violating its secrecy, which, like their treatment of McClelland merely compounds the suspicion that their statements are true. McKinnon's story is essentially like that of Clark McClelland.

Hall suggests that Air Force generals thought the Tall Whites were near-enemies of the Greys. Hall: "I am quite certain that the Tall Whites and the Short Greys hate each other. I am quite certain that the Tall Whites would never permit the Short Greys to come anywhere near their base areas or near to their housing areas or anywhere that their children might be playing, etc." But what is Hall's basis for that assumption asked LoBuono. Hall replied. "I was with the Tall Whites for over two years. Various remarks were made, and in particular, the Teacher (a Tall White) made the point quite clear to me."

Illustration 43 below was done from a detailed description by Hall of one of the Tall Whites with whom he was acquainted at Nellis AFB in the mid-1950s.

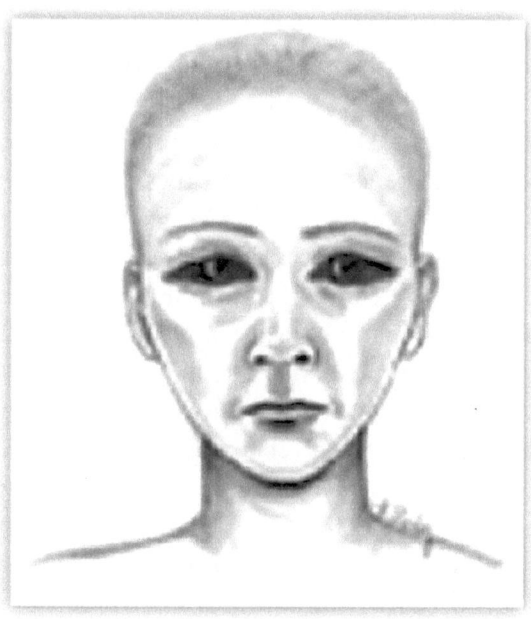

Illustration 43: Sketch of Tall White from description by Charles Hall. Credit: Charles Hall. *Millennium*.

Tall Whites look *too much* like humans, says LoBuono. They appear to be a genetically designed hybrid race placed here for a particular reason. He says that Hall claims he saw a Tall White who wore sunglasses and was able to mix in with a Nevada casino crowd.

At present, according to LoBuono, the literature on aliens shows that aliens evolve from different animal species on planets of different sizes under varying conditions, just as we are evolved from other primates such as the great apes. Aliens reportedly tend to be bipedal but have varying skin color, eye shape, skin characteristics and numbers of fingers. In other words, they look very alien. So, says LoBuono, Tall Whites, who have no independent story about their evolution, may have been created by a group of aliens that took humans from this planet, and then hybridized them to create a headstrong, Tall White version.

But an interesting endnote (no. 18) in the LoBuono article may hold more water. It quotes one Bob Hieronimus's interview with Robert Dean, March 24, 1996 online.[214] He cited the case of David Jacobs, Ph.D., Associate Professor of history at Temple University, and author of several books on the subject of abductees. Jacobs is quoted as saying that human looking "Nordic" aliens were involved in the abductions of some of his patients. Jacobs further says that, "The evidence clearly suggests that the Nordics are most probably adult hybrids of human/alien mating." Further, he references Jenny Randles, an abduction researcher, quoted earlier in this volume, who, at an MIT conference on abductions, said her research showed that in Britain Nordics were responsible for 35% of abductions; in the United States, 6%; and in Europe 25%.[215]

Mention has been made of Robert Dean. He is a retired Master Sergeant in the United States Army. In his twenty-seven-year military career, he worked in the intelligence division for the Supreme Commander of the Headquarters of Allied Powers in Europe, where he had access to top secret documents. Among those were some revealing substantially that at the time, there were just four groups of extraterrestrials that they knew of for certain, and there was one group that looked exactly like us, so much so that it really drove the admirals and the generals crazy because they were sure that they had seen them repeatedly and had contact with them.

Dean also claims that another group consisted of very large humanoids, six to eight maybe sometimes nine feet tall. They were very pale, very white, and didn't have any hair on their bodies at all. This certainly seems to refer to the Nordics. Most likely both these tall individuals and the previously mentioned group, who looked so human, were both Nordics. Some of our anthropologists, as described earlier in this essay, might be "splitters" who take any small difference to define separate species. The "Lumpers" would see these differences as variations within one species. Even six to 6 ½ feet would not seem unusually tall to humans; even 7' though highly unusual, might not make anyone, even a ufologist, think that he, or even she, was an ET.

CHAPTER XXII

STEVE WILSON

We turn to the case of Steven Wilson,[216] a former air force officer, now deceased. He was born in 1933 and died of cancer in 1997. It seems that many whistleblowers have died prematurely from cancer. Whether there is a causal connection is anyone's guess. His military history is mostly according to his own telling, though the absence of any but the slimmest Air Force record is not in itself necessarily significant evidence of unreliability on his part. He claimed he was ultimately trained as a pilot, and saw action as a fighter pilot in Korea. According to him, his tour there included capture by the North Koreans, torture, and escape. He was promoted to captain and made commander of a fighter squadron later ordered to Wright-Patterson AFB. Wilson had been told that he would be assigned to Majestic-12 as soon as this mission was accomplished, and upon his transfer, he would be promoted to major.[217]

Majestic-12, code name Majic12, or MJ-12, he would soon learn, was the super-secret organization which controlled UFO surveillance and interdictions, retrievals and analysis of recovered extraterrestrial spacecraft and occupants, and issues of public access to any information about these matters. Skeptics seem to enjoy denying that such a group ever existed.

After returning to Wright-Patterson AFB and being commissioned as major, Wilson was informed he was receiving top-secret clearance. His indoctrination into the UFO secrets kept by the Majestic-12 then began. He read reports about the crash of a UFO at Roswell, New Mexico, and how Majestic-12 covered it up by putting all documents and expenses with a Socorro, New Mexico crash. As part of his duties with MJ-12, he was assigned to the 1st Special Forces Air Command and underwent special training with Delta Force and then the Black Berets.

Wilson was to later write, "I looked at these fellows I was to train with. Everyone was a trained killer and assassin. But it still didn't prepare me for the MIB [Men in Black, another group that "didn't exist"], the Wackenhuts (private security firm operatives with government covert projects contracts) and all the black operations that exist deep within our government. This was when I was told that I would cease to exist." Major Wilson explained that he was informed that his job was so secret, that stops would be placed on all his records and whereabouts, and that they would be moved to Majestic-12. He was told that he was needed for something very special, and that he would be on a "need-to-know" basis, at least until he had a high enough security clearance.

Wilson recalls, "I felt I was serving my country. I knew nothing of the greed and power of a few men, who were the top executives of Majestic-12." In the summer of 1972, he was assigned to the 1st Special Forces Air Command, Vandenberg Air Force Base, when a man entered his room, flashed CIA identification, and told the major that there was a plane waiting. In twenty minutes, the plane was taxiing to a takeoff with Wilson aboard.

The plane set down on Papoose Dry Lake, a dry lake bed deep within the Nellis Air Force Range in central Nevada, the same base where Charles Hall had served several years in the mid-1960s. Wilson, with his CIA companion, walked to a rock outcropping, on the other side of which, nestled between some large rocks, was an iron door with no handle. The CIA man opened it. They went inside and down a tunnel. At the end of the passageway, Wilson glanced around quickly and still marvels at the size of the structure. "I could swear that the whole damned mountain was hollow. Right down the middle was a runway, and at the end huge doors, that I later found could be opened to allow a plane to take off right out of the mountain." The two men proceeded to an elevator that descended so fast he "almost lost my dinner." Wilson was ushered into an office down the hall to meet the colonel in charge, whose beady eyes, to Wilson had a mean look and a cold and harsh attitude.

Wilson was shown the means, rather exotic, of access and egress from the base. He was also warned that anything he saw was top secret, and that if he so much as breathed wrong or opened his mouth about anything he saw, it would be his last breath. Wilson later acknowledged, "I believed him."

One morning a Lieutenant Colonel Bennet came in. He asked Wilson if he was busy ("Like he gave a damn," Wilson recalled) and said, "Let's go." Wilson followed him two stories down at the super-secret "S-4" UFO technology area. There he saw eight different kinds of UFOs and people all over the area whom he guessed were scientists. The colonel and the major went into a cubicle where there were about twenty officers and civilians sitting around. Wilson looked at Bennet, who correctly read the questioning look on Wilson's face, which he answered with a curt "Forget it."

This much is background, important though it is to an understanding of the degree of secrecy involved, and the credibility of Wilson. We get now to the crux of the narrative, at least from our narrow point of view.

Wilson was stunned, when a woman came in who was at least eight feet tall. She wore a strange-looking jump suit, and there seemed to be not an ounce of excess fat on her body. In a later interview he added that she wore a strange-looking jump suit, which had a "HI" pattern on the right side above the breast line. Wilson narrated some the details of the encounter: She had finely chiseled features, long blonde hair, and the bluest blue eyes he had ever seen. She set a large crystal on the table, and without warning, her fingers began to glow as she ran them over this crystal. A 3-D hologram began to form above it.

Holograms are fully three dimensional portrayals that appear to be actual physical scenes. One can walk around them and see the scene, or object, from different viewpoints or perspectives. But put your hand on one and it goes through like air; solid as it seems, it is only light. A number of these, of similar content, were shown to an abductee named Jim Sparks, who later described historical scenes shown in considerable detail.[218]

Wilson said, "I looked around the room and everyone's mouth was hanging open, and suddenly I noticed mine was, too. Little did I realize that at that moment my life would forever be changed. My past teachings slipped from me as I stared. My whole concept of life did a 180-degree turn, as I watched the hologram, complete with sound, unfold the mysteries of the past and the present and of other worlds."

Wilson related that among the scenes the crystal hologram displayed for the group, was the history of the Earth and of extraterrestrial involvement with it. She also showed scenes from inhabited planets of other star systems. Wilson felt transformed. "When it was over, I knew that, whatever part I was to play in all of this, my life as I knew it had ended forever." There have been a number of other accounts by abductees and other witnesses to the use by ETs of Holograms to demonstrate their narratives.

Wilson became executive officer of "Project Pounce." Created in the final days of December 1980, Project Pounce was an elite group of Air Force Black Berets and military scientists who cordon off the area of any UFO crashes. They retrieve the extraterrestrial spacecraft and any occupants, then "sanitize" the site back to its pre-crash appearance, and intimidate any outside witnesses into silence.

He rose to the rank of colonel and received an ultra-top secret, Cosmic Q, level-27 security clearance, and learned much about the inner workings of the Majestic-12 agency. Wilson's UFO-secrecy duties included interacting with covert "MIB" (men in black) enforcers from the Wackenhut private security firm, whom he came to despise as killers. He also came to know some things about the top command of MJ-12, known as MAJI, including the identity of two of its executive board members, Chairman Henry Kissinger and advisory scientist Edward Teller, both of whom hold the topmost level-33 security clearances. It eventually began to sicken him. He discovered that the MAJI were "so powerful that they acted as though they were above the president and the laws of nature and mankind."

The names of those two prominent and respectable people. Kissinger and Teller, may, to some, seem oddly out of place in a narrative of this nature, but there is credible corroboration. Perhaps the most persuasive is a document published in 2004[219], about 8 years after the death of Wilson.

A group known as "Above Top Secret," has been engaged in attempting to ascertain the truth behind the UFO/extraterrestrial phenomena and the secret working of the government in shielding the facts from the public. This group's attention was soon focused on an unacknowledged subcommittee of the National Security Council with the name of Special Studies Group.

In the 2004 document referred to, it was stated that Henry Kissinger was the head of the Special Studies Group, which was in charge of UFO/ET matters. Edward Teller, the father of the H-bomb, was named as another member. The code name for this group was Aviary. It was said to be in operation since 1947 when it was organized by President Truman and called MJ-12. Kissinger was known as MJ-1. Dr. Michael Wolf, scientific consultant to this group was among those mentioned as having given details to the Above Top Secret group. Still another mentioned was Steven Wilson: "Col. Steve Wilson, USAF (ret.) revealed that he was in charge of Project Pounce, the unit tasked to retrieve downed UFOs and prevent civilian access to them. He also revealed the designations and manufacturers of U.S. antigravity."

Further corroboration as to Teller comes from testimony of Bill Uhouse.[220] He narrated to author Steven M. Greer his career first, for ten years as a Marine Corps fighter pilot, and for thirty years working for defense contractors as an engineer of antigravity propulsion systems. He told of a re-engineered craft that had crashed in Arizona. The craft was taken to recently completed Area 51, and four ETs were taken to Los Alamos. Uhouse said that they could speak understandable English but at Los Alamos only one of them did. He met with one alien they called J-ROD.

Uhouse then said, "The alien used to come in with [Dr. Edward] Teller, and some of the other guys, occasionally, to handle questions that maybe we'd have.

Finally, disgusted with the "unconstitutional and unethical activities" of the Majestic-12 agency and of his involvement in what he called "one of the most dastardly and heinous cover-ups the world has ever known," Wilson left. At retirement, after forty years in the air force, he was flight commander of the First Special Forces Air Command, Vandenberg Air Force Base. The list of decorations he says he received is impressive. After musing over his distasteful experiences for fifteen years, he decided to risk his life and tell all. His many disclosures of sensitive information have been placed on the Skywatch Web page. He was a frequent communicator on the UFO information news groups he founded, as skywatch_ok@msn.com. Stricken with cancer that took his life at age sixty-four, Steve Wilson assessed the price of his years in the "black world" of the UFO cover-up.

"I have no feelings, truthfully. My association with MAJI has left me dead inside. I feel myself still cold and calculating. I never let anyone get close to me. I feel like a human robot. I have killed mercilessly and lied for the good of the country, or so I believed at the time. The things I have seen are beyond human understanding and totally unbelievable. I only have a desire to help humanity somehow through what is bound to come soon."[221]

He was neither the first, nor the last to mention the startling presence of the human looking ETs.

CHAPTER XXIII

JULIO FERNANDEZ

Another was Julio Fernandez. His case was investigated by Antonio Ribera, an experienced and distinguished investigator and author.

Julio, a 30 year old Spanish businessman, was given a battery of psychological tests by reputable psychologists at Ribera's request. Before his abduction he had held no interest in UFOs or parapsychology. He seems to have been an unusual individual. He was a Tae Kwon Do black belt, a photographer, mountaineer, explorer and hunter, hunting being his main passion.

According to Ribera, the tests revealed that "[He had] an IQ higher than normal, together with a very well balanced personality, perfectly integrated and without a trace of the psychopathetic. Julio is not a mental defective nor a concoctor of tales, nor a mythomaniac. On the contrary, he is a very realistic person, very objective, and, above all, incapable of lying."

Ribera adds that the psychologist, Jordán Peña, who, because of the depths of his experience, would be impossible to deceive. It is apparent also that Julio was a remarkably perceptive observer, even under very tense circumstances. He had admitted to investigators that he was frightened by the episode.

PALE SKIN, GIANTS, AND THE GREAT TRANSITION

It was on February 5, 1978 that Julio's skepticism about extraterrestrials ended. On that day he left his home near Guadalajara about 3:30 AM to go hunting. He was accompanied by his dog Rus. When his car unaccountably stopped, Mus growled, so Julio, as a precaution, loaded his Winchester with five cartridges. He related that he got out of his car and was shocked to see a couple of humanoids coming toward him on the road, some creatures obviously not of this world. Julio later said to investigators, emphasizing his disinterest in UFOs or extraterrestrials, "It was like Karl Marx beholding God." The place where he alighted from his car was not at a place he had intended. He got out because his car engine and audio cassette player had both quit working.[222]

He had, in fact, never intended to be at that spot. For some reason for which he could not account, he took a road that led north by northeast, the opposite direction from which he intended. Also for some reason he suddenly felt that he had to be at a certain location fifteen kilometers further on and pushed his foot onto the accelerator. Under hypnosis he recalled that he suddenly braked and stopped the car for no apparent reason, and the car, on its own accord, backed onto a dirt road. It was now about 4:30 AM. That is when the engine and cassette player quit, whereupon he was approached by the humanoids. Unintended deviation of this nature has been experienced by many other ET abductees.

Julio described the craft to which he was brought as an awesome sight, shaped like an inverted soup plate, with a silvery, metallic color. He was in the craft until about noon, a total of about 7 ½ hours, and gave the investigators a detailed and very interesting description of the interior of the UFO. He also related a number of highly interesting episodes and exchanges, through mental telepathy that occurred during his time aboard the craft. Interesting though they are, they are not of primary relevance to our thesis.

We turn to his description of the humanoids: They had. according to Julio, extremely broad shoulders, with powerful dorsal muscles showing prominently, narrow waists, and they seemed athletic. Differentiating them from humans of Earth were exceptionally long arms and hands, large crania and very large eyes. When they communicated with him, Julio thought they were speaking, but soon noticed that their lips did not move, and he then assumed they mentally communicated with him.

When they asked him to "calm down" and follow them, he did so with his rifle and his dog. He later said they walked with a majestic, elegant, and rhythmical gait.

The skin color of the aliens was described as "extremely bluish-white, typical of those who seldom go out in the sun, giving them a "Nordic" caste. Whether it was Ribera or Julio who used the term "Nordic" is not certain from the report. It was Julio however who stated that because they could walk so gracefully and rapidly in the dark, he believed their night vision was superior to ours. He stated he had to take an average of one and a half strides to keep up with their one.

The assisting physician to Ribera, a Doctor Pérez, suggested they came from a habitat where the light is soft and mellow, or perhaps does not hurt the eyes. He stated that this fits in with the absence of any eye lashes, the function of which is to serve as shades or awnings, and with the low amount of pigmentation in the iris, and also with the parchment-like color of the skin.

As summarized by Ribera, Julio further described in more minute detail the physiognomy of the captors:

> The extremely long hands looked …. feeble and bony, very fragile, like the hands of pianists. Their immensely long, fine fingers were thin and knotted—there was nothing but tendon and bone to be seen beneath the skin and their fingernails were short and clean and normal… Their heads were also different from the human head. The forehead went straight up for a good distance, and then into a great high curve; it protruded more than our heads do, and was also much bigger. One of their most typical features was a prominence over the eyes, very massive… As far as the area of the temples, the entities had the parietal bone very much developed, their size and degree of convexity being considerable, the head almost a huge globe. He likewise observed no ears, these being presumably covered by the hoods. There were no eyebrows, eyelashes or traces of beard or hair

PALE SKIN, GIANTS, AND THE GREAT TRANSITION

Further, reported Julio, the face was bony with a long thin nose and high cheek bones. The mouth showed as a mere streak, likewise very thin, in place of lips. Among outstanding features was the "cone shaped chin," described as "enormous, projecting outwards and downwards and ending in a point."

Ribera also summarized Julio's description of the eyes, "like two huge beacons projecting from the face... the eyelids were oval in contour and did not terminate in an angle or fold, as do humans. The iris, gigantic, was double the size of the normal iris of a human eye, and was of a pale blue, almost transparent, shade. The pupil seemed to be extremely dilated, and this gave them a hypnotic gaze as though in a kind of permanent state of fright or shock." Ribera stated though that paradoxically it had a tranquilizing effect.

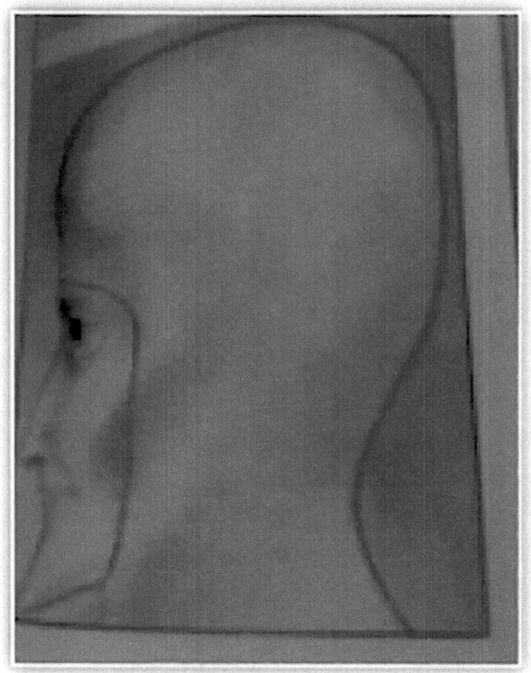

Illustration 44: Profile of captor of Julio Fernandez, rendered from his description.
Credit: Timothy Good: *Unearthly Disclosure* and Maite Alvarez López

Illustration 45: Frontal View of Nordic Captor of Julio Fernandez, rendered from his description.

Credit: See illustration 44.

No doubt these entities could never pass as Earthlings, and there seem some significant differences from descriptions of ETs more obviously Nordic. I include this report as it more closely resembles the descriptions of the Nordics than those of any other race.

> He also described their clothing: They wore seamless one-piece pastel-green overalls that reached to the feet and gave off a very faint luminosity, with no zippers nor openings and gathered at the waists... The overall was sufficiently tight fitting for their muscles to show... No fibres, no thread, and no designs were detected in it. The material was smooth and did not rustle when they moved. Their heads and shoulders were covered by pale yellow covers or hoods, leaving only the faces uncovered... Their hoods, like their five-fingered gloves... were made of

a texture resembling satin-stich, very fine and fitting the parts of the body closely.

Of primary interest to us is the fact that during an extended physical examination, his captors took a sample of semen from him. This points to a high probability that they, like the Greys, still have an agenda of hybridization, which may be a source of the tension between the two extraterrestrial races.

This conclusion may be further fortified by the fact that the cosmonauts, as Julio called them, communicated to him telepathically, twice, once during and once after, his examination, that other shorter beings were coming here who were less evolved ethically. It was stated that this other group is engaged in probing and programing the minds of humans whom they have contacted or kidnapped... and that relations between the two groups were not precisely the best. Ribera's reaction: "Yet again, a conflict of interests among alien species regarding Earth is indicated."

Julio's later drawing of one of his captors, (Illus. 44 and 45) done at the request of an investigator was clearly of a Nordic and not of one of the Grays. Though obviously not knowing, nor caring about the various groups of ETs, he found the bluish white skin worth mentioning, and it was probably he who used the term "Nordic caste." He made no mention of height, from which we must conclude that there was nothing striking about it. In that connection, we must bear in mind the different characteristics of even the most tightly knit groups of earthlings. We must also consider the possibility that hybridization may well have existed between different races of ETs. Some of the Greys, very small creatures, who work in the UFOs doing service in the removal and implantation of embryos, have reportedly been supervised by significantly taller individuals who seem to be in charge, though in all probability, from descriptions, they are not Nordics. The pervasive view of tension between the two groups would also seem to militate against it.

Asked by Julio why they, the "cosmonauts" did not seek out eminent scientists of the Earth with whom to communicate, they responded that the greatest of the Earth scientists did not come to the level of their most "modest technicians." What they were seeking was the "warm human qualities" that had atrophied in their race, lost throughout the course of centuries of a harsh and difficult self-imposed evolution in a cold and hostile environment. As Julio put it to investigators, they didn't have a Beethoven. Ribera, author of several books on the subject of extraterrestrials and an investigator of this case and others like it, was quoted by author Timothy Good as saying, "In us they see—and admire—what they had centuries ago and they have now irremediably lost... they are superb biological machines, frigid and perfect, but robotized, retaining merely an archaic sense of humor as a relic of the lost humanity." Whether he was speaking only about Greys or whether Nordics were included is not certain.

Timothy Good also mentions one Ed Walters, one of a number of humans captured by the other race of aliens mentioned, who claim that those aliens are likewise fascinated by human emotions. Walters claimed that during his abduction, they clamped a headset on him and stimulated his memories so that the various emotions engendered could be transferred directly to the aliens.

That episode is an eloquent description of the faith of that alien group in technology. Even the stimulation of long-lost emotions in themselves, they hope to recover though a reading of the electrophysical activity in the human brain upon the experience of certain emotions. No one can say they are wrong in doing so. About whether the emotion causes the electrical activity or the activity causes the emotions is still debated by our neurologists, not to mention our philosophers. The aliens may know more than we do, or they may have developed a view of the entire emotional structure, warped in the direction of technology.

CHAPTER XXIV

EISENHOWER AND THE NORDICS

There is much evidence that on February 20th 1954 President Dwight Eisenhower met with a delegation of what were probably Nordics at Edwards Air Force Base in California, then called Muroc AFB. It was in fact the first of at least two meetings with ETs, the second taking place on February 9th 1955 with Greys at Holloman Air Force Base in New Mexico. The first meeting, with Nordics resulted in no agreement between the parties. The second, with the Greys, did result in an agreement, called the Greada Treaty that was short lived. Our prime interest is with the first meeting and with the fact that it was undoubtedly with Nordics, and we should therefore look at that with some detail.

We start with the fact that sometime during his first year in office, Eisenhower received a document prepared by a governmental department briefing him on the current situation and history of UFOs on our planet. It advised him that if he wanted to meet with representatives of the aliens, it could be arranged. He was further advised that security would not be a problem as the aliens had shown no hostility toward us, nor any inclination to do us harm.

This information was stated publicly in a speech by a former state representative from New Hampshire, Henry McElroy in May 2010 after he was no longer in office. He had served on committees dealing with federal and state relationships, and had access to many documents of a classified nature. McElroy stated that though he could not prove it, he believed that Eisenhower in fact had met with aliens. Much evidence indicates that he was entirely correct. The following is a summary of reports from researchers on the subject, primarily, William Moore, Michael Salla and Timothy Good.[223]

On February 16, 1954, Eisenhower announced that he was going on a golfing vacation to Palm Springs, California. This was somewhat surprising as he had only recently returned from a quail shooting vacation in Georgia. The next day he did fly to Palm Springs where he stayed with a friend at the "Smoke Tree Ranch," owned at the time by a friend, Roy Helms, located about 100 miles from Muroc. He was there from the 17th through the 24th of the month. But on the night of Saturday the 20th for a number of hours, he turned up missing. This drove to distraction the members of the press that were following his every move and hanging on his every word during those tense days of the cold war. The inevitable rumors started to fly: Ike was sick, and at one point, there were guesses that maybe Ike had died.

James Haggerty, the press secretary, called a quick press conference and assured the assembled newsmen that there was no cause for alarm. The president had been eating fried chicken and broke the cap on a tooth, and had to be rushed to a local dentist for immediate repair work. Big relief all around. The next night, Sunday, there was a steak fry with many guests. A guest specially invited by Eisenhower was a dentist, Dr. Francis Purcell. He was introduced to all of the guests as "the dentist who treated the president."

So far so good. There were attempts by member of the press to interview the dentist, but he refused to discuss it. In 1974 Dr. Purcell died. Someone, either a researcher or a family member, searched his record looking for the records of his treatment of the President of the United States. The records were intact, including those for the time period in question, but there was no record of his having treated the president in February 1954 or any other time. In the late 1980s, probably motivated by increasing evidence of a meeting with aliens in that time frame, a researcher, and columnist, William Moore, visited the Eisenhower Library in Kansas City where the deceased president's medical records are kept.

The files and the indices were voluminous, but contained nothing showing dental treatment by Dr. Purcell at any time, nor treatment for dental work by anyone during February 1954. There was a separate file marked "dentists," but Dr. Purcell's name was not among those listed. There was however a file of thank you notes to everyone contributing to his stay in Palm Springs. These included everyone who met Ike's plane on arrival, everyone who offered to play golf with the president, everyone who sent flowers, and to the minister who conducted the religious service that Ike attended on Sunday morning. But there was nothing to the dentist who supposedly was called out on a Saturday night from home to treat this broken cap on Ike's tooth. So Moore talked to the widow of Dr. Purcell. She remembered almost everything about the steak fry, apparently including who she met, where it was, when it was, and many details. But she remembered nothing about her husband having been called upon to treat the President of the United States the previous evening or any other time.

Admittedly, none of this proves that Eisenhower met with aliens that night, but it seems to prove that the story about the broken cap did not hold water; it was not true and was obviously a cover story for something else. But what?

Let us go back to early 1954. On April 16th, a gentleman named Gerald Light wrote a letter to a friend, one, Meade Lane.[224] Light was known to be a student of UFOs, a subject that did not interest many at that time, and generally precipitated even more ridicule than it does today. That he was also interested in the metaphysical, or paranormal, did nothing to alleviate the ridicule. But he was nonetheless respected by many of a scholarly bent and considered a community leader, even if those he led were likewise not taken very seriously by most of the public. In the letter to Meade Lane he said that he had just returned from Muroc AFB where he was one of a group of advisors to President Eisenhower at a meeting with extraterrestrials. He named others of the president's entourage as including Franklin Adams of the Hearst newspapers, Edwin

Nourse of the Brookings Institute, economic advisors to both Presidents Harry Truman and President Eisenhower; and Cardinal James MacIntyre, head of the Catholic Church in Los Angeles.

There is considerable corroboration. Light mentioned that he and the others of the president's group were permitted, with the consent of the Nordics, to examine 5 UFOs that were in a hangar. According to a well-respected UFO researcher, Timothy Good, a retired Air Force test pilot stated that he had seen the five UFOs under guard in the hangar, and described two as cigar shaped, and three as disks, both shapes being often seen in our skies. He also said that airmen returning from pass were not permitted back on the base. Michael Salla adds that a Los Angeles Times reporter, hearing of interesting happenings tried to enter the base, first by plane, then by car, but was turned away both times. He referred to this as "unprecedented," the base always having been open to the public.[225]

The meeting had reportedly been requested by the aliens, the Nordics, and their avowed purpose was to start an educational program with the people of the Earth. They were seeking to make their presence known and to explain their identity and the interplanetary subject generally. If so, the assembled group of advisors was well suited to give advice to the president as to the impact such a disclosure would make in their respective areas, including the economic, the public relations, and religious problems.

Light's letter had a note of triumph concerning his long acceptance of the UFO and ET phenomena. He spoke of the emotional and mental pandemonium that was occurring among the scientific community, presumably among some of Eisenhower's close advisors. Considering that it involved so many individuals, the pandemonium must have occurred among these scientists after the meeting of February 20th. Light expressed the opinion that the president would ignore the conflict between the various 'authorities' (the quote marks are those of Light) and go directly to the people. "From what I could gather," he continued, a statement was being prepared for delivery about the middle of May.

There are a few details added by Timothy Good, author of *Need to Know*.[226] We turn now to some of the portion of his book relating to this meeting. Much of it is hearsay several times removed, but still, the personnel links considered, the reports seem beyond the realm of common gossip.

He says, as reported by Desmond Leslie, an author and former pilot with the Royal Air Force, that he, Leslie, learned from a USAF officer that a disk, estimated at one hundred feet in diameter, landed at nearby Muroc AFB. The American UFO researcher, Gabriel Green, was later told by a military officer that he witnessed the overhead arrival of five UFOs while he was engaged in firing practice at Muroc under the command of a general. The general ordered all antiaircraft batteries to fire at them, but the firing had no effect. The craft landed near a large hangar.

Author Brinsley Le Poer Trench, sometimes known as Lord Clancarty, later spoke to a retired colonel who had served as a USAF test pilot. Clancarty told a reporter that he, Clancarty, had been told by the colonel that two of the alien craft that landed at the base were cigar shaped and three were saucer shaped. Whether the colonel was the same officer who described the UFOs to Gabriel Greene, we do not know and now could not find out. An immediate security blanket was thrown on all of Muroc.

The crafts were placed in a hangar under guard, and men returning from leave were not allowed back on base. Two other witnesses, Don Johnson and Paul Umbrello, also reported having witnessed one of the craft near Muroc on the evening of February 20.[227]

The colonel to whom Clancarty spoke claimed to have been one of six persons present at a meeting between crew members of one of the craft and President Eisenhower who had been spirited away from Palm Springs for that purpose. The meeting had apparently been requested by the alien crew. Lord Clancarty told a reporter that the aliens, according to what he had been told, "looked human-like, but not exactly." They had the same proportions as humans and were able to breathe in our atmosphere. They spoke English, but did not say where they came from.

It is doubtful that most of the creatures we have seen thus far, except for the Nordics, would be described as "human-like, but not exactly." The report that they spoke English is at odds with other accounts of telepathy, however not everyone comes away from such encounters aware that the mouths of the ETs never moved.

Others present at the meeting besides the colonel were the same individuals identified by Gerald Light. The colonel may be Good's source.

Good also agrees that evidence supports the fact that on February 20, Ike disappeared, and failed to show up for a press conference, causing pandemonium, and that the official explanation was that the president had gone to a local dentist caused by the loss of a cap on a tooth. Good also states, as does Moore, that no evidence for the trip to the dentist, such as a White House memo, has ever surfaced.

The aliens conveyed to Eisenhower the relationship that they wanted to start with the people of the Earth. According to Good, Eisenhower was said to have been ill at ease, but replied that he did not think the Earth was ready for that and that such a revelation would create a difficult situation for everybody. The colonel said to a reporter that the aliens seemed to understand and indicated that they would continue to make isolated contacts with humans. Reports from several of the whistleblowers are to the effect that the U.S. representatives wanted access to their technology and that the ETs refused unless we agreed to dismantle our atomic arsenal. This was more creditably the reason for our rejection of the idea. We wanted above all their technology. The dispute between the military people is understandable. With the ET technology we could possibly be impervious to any attack. But whether the technology without an atomic arsenal was more or less advantageous than the arsenal without the technology could well be a close call.

The meeting was followed by a demonstration of their ability to overcome gravity and to make their craft invisible. The colonel said that this disturbed the president greatly "because now none of us could see them although we knew they were there." There is some further evidence of productive meetings involving exchange of some technological assistance from the aliens, or perhaps a few volunteers acting on their own. The detailed circumstances are compelling and are corroborated by other evidence.

But as far as anyone knows, no agreement with the Nordics was ever reached, except perhaps with a few of the alien individuals, nor was any announcement ever made. The nature of the discussions and the reasons for the failure to reach an agreement are based on statements of some close to the discussions who ultimately became whistle blowers.

A year later came a treaty, known as "Greada," reached according to the whistleblowers with the Greys at Holloman AFB in Arizona. The meeting, the nature of the agreement, and its quick collapse, are interesting subjects, but extraneous to our interest in the Nordics. Some of the whistleblowers who reported that controversy were the same who told the story of the Muroc meeting. Who were they?

One was Don Phillips. He is a former air force serviceman and employee on clandestine aviation projects. According to him, in a statement to Steven Greer in 2000:

> We have records from 1954 that there were meetings between our own leaders of this country and ET's here in California. And, as I understand it from the written documentation, we were asked if we would allow them to be here and do research. I have read that our reply was, well, how can we stop you, you are so advanced? And, I will say by this camera and this sound, that it was President Eisenhower that had this meeting. And, it was on film, sort of like what we are doing now. Bringing it up to date, the NATO report gave that there were 12 races. To make a final summary, they had to have contacts to go to these races in order to understand who they are, what they are doing, and what they could do. Are these ET people hostile? Well, if they are, with their weaponry they could have destroyed us a long time ago, or could have done some damage.[228]

Another was Charles L. Suggs, a retired Sergeant from the US Marine Corps.[229] His father, also named Charles L. Suggs, served as a Commander with the US Navy, and with others attended the meeting at Muroc AFB in 1954. He died in 1987. According to the younger Suggs, narrating no doubt what he heard from his father, Ike's group met and spoke with two white haired Nordics that had pale blue eyes and colorless lips. One, the spokesman, stood a few feet away from the president and would not let him approach any closer. Suggs said that a second Nordic stood on the extended ramp of a bi-convex saucer that stood on tripod landing gear on the landing strip. The aliens said they came from another solar system, and posed detailed questions about our nuclear testing.

Milton William Cooper was a member of the U.S. Naval intelligence briefing team for the commander of the Pacific Fleet, between 1970 and 1973, and had access to classified documents that he had to review in order to fulfill his briefing duties. During his military service, he claims to have witnessed naval documents describing the recent history of extraterrestrial life on Earth and how extraterrestrials have interacted with various government and corporate entities.

His credibility has been questioned not only by perennial skeptics and debunkers, but most seriously by Budd Hopkins, a well respected researcher and author whose own credibility appears beyond reproach, despite attacks by the usual debunkers. Hopkins died in August 2011. He had painted a picture of Milton Cooper as a man unhinged, which external circumstances seem to corroborate. Hopkins detailed very vicious personal attacks and threats on himself and others, which clearly render Cooper unreliable, except as he may be corroborated by others of more substantial integrity.

In 1998, Cooper was living in Arizona. He was wanted on charges of tax evasion and was the subject of an arrest warrant for "aggravated assault with a deadly weapon" against a local doctor. Just before midnight on November 6, 2001, officers of the Apache County Sheriff's office converged on Cooper's home to arrest him on a warrant arising from the threat complaint. Cooper was considered armed and dangerous. During the execution of this warrant, Cooper shot and struck a sheriff's deputy and was then fatally shot by another deputy.

The following summaries of his account nonetheless appear reasonably stated and, hence, may predate his mental collapse. There is nothing to hint of illusory or hallucinatory thinking, unless one is persuaded to dismiss all such accounts in similar fashion. It is difficult to tell approximately when Cooper's mental breakdown occurred or to exclude the possibility that at one time, now uncertain, he was more in control of his faculties. He is cited here for whatever facts he may give us that seem corroborative of others and which others, in turn, are corroborative of his.

A detailed summary of his claims is described by Salla, who quotes Cooper substantially as follows:

> In 1953, Astronomers discovered large objects in space which were moving toward the Earth. This was four years before the first man made (Russian) satellite. It was first believed that they were asteroids. Later evidence proved that the objects could only be Spaceships. Project Sigma intercepted alien radio communications. When the objects reached the Earth they took up a very high orbit around the Equator. There were several huge ships, and their actual intent was unknown. Project Sigma, and a new project. Plato, through radio communications using the computer binary language, was able to arrange a landing that resulted in face to face contact with alien beings from another planet. Project Plato was tasked with establishing diplomatic relations with this race of space aliens.
>
> There is corroboration of the first part of Cooper's narrative involving the fact of a satellite in 1953. It was reported by John Cohane in 1977: "In 1953 new long-range Air Force radar had picked up a huge object six hundred miles out circling the Earth near the equator at a speed approaching 18,000 miles per hour. Not long after, a second one was discovered in orbit about four hundred miles away."[230] That year puts the satellites orbiting within the time frame of the February 1954 meeting at Edwards (Muroc) AFB.

Still further corroboration comes from an article in *Newsweek* magazine of July 4th 1960, shortly after the first manned flights by both Russian and American craft. The article, entitled "The Strange Intruder," reported:

> A growing number of scientists are now convinced that Spacetrack (National Space Control Center), for all its diligence, may have overlooked at least one space vehicle neither Russian or American, but out of this world—indeed, out of this solar system. This satellite, they suspect, is a visitor sent by the "superior beings" of a community of other stars within our Milky Way galaxy.

In its issue of February 1960. *Time* described it as a "mystery spook satellite."

Returning to the events of February 1954 as related by Cooper, a race of human looking aliens, later termed "Nordics" by military personnel, contacted the U.S. Government. These Nordics warned us against the aliens that were orbiting the Equator and offered to help us with our spiritual development. They demanded as a condition that we dismantle and destroy our nuclear weapons. They refused to exchange technology, claiming that we were spiritually unable to handle even the technology which we then possessed. They believed that we would use any new technology to destroy each other. They stated that we were on a path of self-destruction and that we must stop killing each other, stop polluting the Earth, stop raping its natural resources, and learn to live in harmony.

It was believed by our leaders that the destruction of our nuclear weapons would leave us helpless in the face of an obvious alien threat, not to mention our adversaries on Earth. We also had nothing in history to help with the decision. Nuclear disarmament was not considered to be within the best interest of the United States, and the proposed treaty was rejected.

According to Cooper, as with many others we have cited below, the Grey extraterrestrials who later signed a treaty with us proved to be untrustworthy. What was one of the more interesting aspects of Cooper's account was that he described the Nordics as giving the U.S. officials warnings about the "Greys."[231]

Another "whistleblower" is John Lear, the son of William Lear, the famous creator of the Learjet. John is a former Lockheed L-1011 captain who flew over 150 test aircraft and held 18 world speed records. During the late 1960s, 1970s and early 1980s, he was a contract pilot for the CIA. He was also a close friend of former CIA director William Colby. Lear confirms that there had indeed been a warning from another extraterrestrial race prior to an agreement being eventually signed with the "Grays," and he, like the other witnesses, claimed that in 1954. President Eisenhower met with a representative of an alien species at Muroc Test Center. This alien group suggested that they could help us get lid of the Greys, but Eisenhower turned them down because they offered no technology.

Lear is also said to believe that the government may have made secret deals with the "aliens," referring to the Greys, actually exchanging humans for advanced technological data. Supposedly, the government was to be provided with a list of those being abducted so they could maintain a vigil over them after their experience and make sure that they were not being harmed in any way. Unfortunately, the aliens took advantage of the situation, taking away tens of thousands and implanting small transmitters inside their brains which can be activated for some sinister "mission" at some prearranged future moment.[232]

In his statements about agreeing to abductions and implanting transmitters, Lear is corroborated by one Phil Schneider. This subject, of course, refers to the treaty with the Greys, but being part of Lear's testimony, corroboration, where it exists, seems in order. Phil Schneider is a former geological engineer who was employed by corporations that contracted to build underground bases. He worked extensively on "black" projects involving extraterrestrials. "Black" refers to super-secret organizations and projects for which money is appropriated but need not be accounted for. Schneider's knowledge of the treaty would have come from his familiarity with a range of compartmentalized black projects and interaction with other personnel working with extraterrestrials.[233]

The whole subject calls for some discussion and articulation of certain significant questions it raises. The meetings with Eisenhower were in the early 1950s. The first the public knows of alien abductions is a 1957 episode involving one Antonio Villas-Boas in Brazil. But even that was not widely publicized and may not have been known by our government for some significant period. A much publicized abduction of Barney and Betty Hill in New England occurred in 1961, and as far as is known publicly, that was the beginning of any significant awareness of the human abduction phenomenon.

Hence, these statements from Lear and Schneider either bespeak of a much earlier beginning of the abductions than the general public awareness of them, something perhaps known only to the government and an inside few. Or it is indicative that the witnesses' testimony is made of the whole cloth, contrived or imagined by them and without any substance at all.

I would be very, very surprised if it all began suddenly in 1961 or 1957. The aliens have been coming to our planet for many centuries, perhaps hundreds of millennia or more before then. I think that before 1961, the reluctance of abductees to speak openly would be many times that of such victim after the path was smoothed, however little, by Barney and Betty Hill. There were then no such people as Budd Hopkins, Dr. John E. Mack, David Jacobs, Barbara Lamb, or any of the other of the still too few qualified and interested people willing to listen and to help. I think the first hypothesis above to be a more likely one than the second. I see little reason to dismiss the word of these two witnesses, probably unknown to each other and unmotivated to invent such unnecessary detail. Admittedly, the possibility, however, cannot be ruled out. Admittedly also, if the statements were false, it would cast serious doubt on the credibility of the balance of them.

We return to more testimony. Lear's idea of more than one extraterrestrial race interacting with the Eisenhower administration is supported also by former Master Sergeant Robert Dean. In his twenty-seven-year military career, as already mentioned, he worked in the intelligence division for the Supreme Commander of the Headquarters of Allied Powers in Europe, where he had access to top secret documents. Among those were some revealing substantially that at the time, there were just four groups of extraterrestrials that we knew of for certain, and the Greys were one of those groups.

We have also already heard from Dean about the ones who drove the top brass crazy for their close resemblance to us. Of the two other groups, Dean claims that one consisted of very large humanoids, six to eight maybe sometimes nine feet tall. They were very pale, very white, and didn't have any hair on their bodies at all. About them we have already referred to the possibility that Dean is a "splitter," making two races out of one according to height, some like ours, some like giants.

There was another group that had sort of a reptilian quality. Military people and police officers all over the world, according to Dean, have run into such types. They had vertical pupils in their eyes, and their skin seemed to have a quality very much like what you find on the stomach of a lizard. We should remember our earlier mention of Werner Von Braun, the German rocketry genius, working in America after the war, who described as reptilian the corpse of one he had seen while working in the American southwest.[234] Von Braun, whatever his dark history as a Nazi, does not seem to have been one with tendencies to wild imagination. Many qualified scientists believe that except for him and his devotion to fact, we would not have reached the Moon when we did.

There are differences in the accounts of the different researchers, and in the different witnesses, not unusual or surprising to those who have been, for any length of time, engaged in investigative work, particularly concerning events that are decades old. All circumstances considered, there is, in truth, remarkable congruence of the important factors. It all seems to establish without doubt that the Nordics are still among us, and in all probability, mating with some humans of planet Earth. We turn now to a different type of witness about the subject, more important for the numbers of them involved and the ring of truth of their accounts, admittedly a very subjective element, rather than whistleblowers with inside sources. It is good that we have both.

CHAPTER XXV

WORLEY AND HIS AND OTHER WITNESSES

It happens that in the United Kingdom, as in the United States, the Nordics have been considered the most human in appearance of all extraterrestrials. They have been described in the Kingdom as having long blond, almost platinum hair and "amazing" blue eyes, and very beautiful. The males are said to have muscular physiques. They are likewise said to be the extraterrestrials that most often reproduce with human beings, mostly their men with our women. Two women, Elizabeth Klarer and Cynthia Appleton have claimed to have had sexual relationships and borne children with Nordics.[235]

When Don Worley embarked upon his research of the Nordic-types of extraterrestrial, he found that many of those claiming encounters and abductions were describing the same types of beings and events without knowing each other.[236] Most often those with whom he spoke were lifelong experiencers. Most frequently their contacts occurred during the night. In rare instances however, individuals have said that their encounters have transpired in broad daylight under like circumstances.

In a total of approximately 100 cases, 36 of the reported episodes involved Nordic-type aliens. Worley discusses various aspects of his conclusions. First is the appearance and nature of the Nordics. Like other investigators of the ET phenomenon, he assigns fictitious names to the experiencers, as they are often called, to protect their identity. Worley assures us that he has full particulars of each one, including correct names. Obviously he cannot vouch personally for the accuracy of any individual's report. Even more obviously, neither can I.

Nonetheless, I am summarizing a little less than half of his individual accounts involving those ETs who appear to be Nordics. The selection was based on what I perceive as being most likely to be truthful objective accounts. This does not mean 100% accuracy in all details, something rarely attainable, but accounts appearing free of roaming imagination, inability to separate fact from fantasy, and absence of a desire to impress the interviewer.

We hear first from "'Donna," a Canadian who holds a BA degree. Her experiences started in childhood while residing in Scotland. She claims to have come upon some Nordics "along with the worker Greys and a group of child-like beings who were very pale." She continued:

> They were all in one place and the Nordics—who were blonde—wear silky smooth, light blue suits. The only symbol of their victims I recall is a dragon-like snake with wings. The males were handsome with bottomless blue eyes, set in angular faces. The only emotion they displayed was serene amusement until they began to mentally talk to me. The place had a beautiful glow and atmosphere of serenity. Most of the beings looked young, but there were some there who looked older with beards and flowing robes. I had an intense longing for these beings after I left them, and an intense love, respect, and awe when I was with them.

Judy, an extensive Texas experiencer recently emotionally declared that if the extraterrestrials who manipulated her into the mess she is in don't help her soon her life is doomed.

> Yes, around the age of 11 or 12, I had the first experience with the Blondes as you call them, the Elohim as I call them, when I was sitting in our yard

near the plum tree. Suddenly my flesh began to itch and I heard a buzzing sound. Directly in front of me, I saw them through a window in the air. They were very, tall, exceedingly handsome, and dressed in White hermits and suits. The one sitting down had on a long white robe and their eyes were weird for they had light coming from them. I felt they were talking to me, but now I don't remember what was said. After the image vanished the itching stopped and I sat there for a very long time. I knew something truly remarkable had happened to me in this chilling experience."

It should be noted that the Nordics have often wrongly been called "Elohim," the Hebrew word for God. More correctly they are called Nephilim, the offspring of God, and a word meaning, among other translations, giants.

Andy Anderson, an Indiana man, has been abducted far too many times for his liking. According to Andy, there also exists a shorter, adolescent appearing Nordic. This kind of entity approached Andy in a normal daylight setting as he worked in a supermarket. The two who approached him spoke a strange language making him think they must be retarded. So began years of frequent abductions in which a Sanskrit-like language was used to communicate. Andy was informed of this and when he spoke the language in public he ended up in a mental hospital under observation. Worley comments that Sanskrit was the language of the ancient gods of India who flew in aerial craft. It has surfaced in modern America in this case through an alleged abductee.

(In the Vedic literature of India, including the *Rig Veda*, the *Mahabharata*, the *Ramayana*, and the *Puranas*, there are many descriptions of flying machines that are generally called Vimanas. They have many features reminiscent of UFOs. They are unstreamlined structures that fly in a mysterious manner and are generally not made by human beings. Whether they are factual descriptions, or imaginative fictions is debated.)[237]

THE GIANTS:

According to Worley, there also exists a giant type of Nordic. On a Mediterranean island the titan sized entities were called Genitori. In South America a Costa Rican contactee was taken to a base in the Andes Mountains run by giant Nordics. There were also Greys and humans there.

Nordic contact, says Worley, has also been reported in such diverse places as England. Spain, Mexico, and Canada. In northern Mexico, in a "zone of silence" near Ceballos. Durango, Nordics have wandered into town from the desert, accepted refreshments, and then vanished back into the desert. In Canada, one large religious group still has contact with Nordics they call "overseers." This intimacy, they say, extends back into ancient days in Europe.

Initially, Worley continues, he was surprised to learn that the Nordics are the controllers of some of the android-like Greys, there are several examples of such cases.

Near Red Fork Falls, Uanaka Mountain, Tennessee, on one August evening in 1990, five persons said they watched a curious event. Near a large fire was an unusually tall naked man with long blonde hair. Dancing around him were some 30 midgets with large heads.

There are some startling things heard from the Nordic experiencers. There are a number of close contact cases in which the persons insist that their Nordic friends are physical in every sense of the word.

Worley speaks of "Special Revealing: Episodes:"

> Janice, resides in New Hampshire. Since a child she has considered the Nordics as her "other family" and perceives them as true physical beings. She insists, "They are not gods or angels. They are another race who are here because they are responsible for us. I have seen them breathe in a physical state, but not in the non-physical state. I have seen them sit, lie down, have sex, but not go to the toilet. However I have used their standup toilet." She also said, "It is the fear a person is living through that changes his or

her perception of the experience. Just because I am not afraid doesn't mean I never was."

Amy lives in Michigan, but is an acquaintance of Janice. Amy has tried to help others. Worley's last communication from her indicated great tension due to the arrival of new ETs who come to find out what the other ETs are doing. The situation soon required her hospitalization. She has not seen the Nordics eat, but has seen them resting on what looked like a chaise lounge. Amy told Worley, "As you know, I believe there are different groups of tall beings making contact. When I stated I had not seen any Nordics with the Zeta's (Grays) maybe what I should have said is that I don't have any recollection of seeing them. I know that Janice and I have the same Zeta [Grey] contact. She also said:

> "Janice does see the tall beings the Pleiadians [a name often used for Nordics], on board with our group of Zetas [Greys]. I would say that not all Nordic types are from the same community and perhaps their goal or agenda is not the same. The Zetas I am in contact with do have humans living among them. They are from genetic material that they have collected from us earthbound experiences. I have met several of their people, both human and Halflings [apparently hybrids]. I have been shown different living areas for these humans on board."

A highly respected pastor from Canada, who is also a physician, says that as a child he was held on the lap of his lifetime mentor, named" Gold" (The entity has a golden sheen to its eyes which can change color according to its mood), and watched what was done to his loved ones. Over 300 lost time periods and many abduction details have been recalled over 45 years. Some of these memories involved groups of persons in his religious group. In his intimate relations with those he calls the "Overseers or non-Terrans," the cleric has seen them breathe, eat, drink, lift some abductees out of a lake with their arms, and glide over the surface of the water.

The pastor at one point told Worley, "Yes, Don, the ones I have been around do respirate. I have touched and been touched by them on plenty of occasions. You don't think that what I felt was real? Thanks to the number of events wherein I've been in the presence of ephemerals. I assume that I've come to know the difference."

In a later conversation, the pastor described some medical instruments and techniques used by his erstwhile benefactor. His description seems similar, at least in many respects, to the insertion of tiny metallic-like objects placed in the noses of young children. In adults insertions of these small objects have often been found in the skull or the limbs. Much speculation describes and equates these objects as a means of keeping track of the whereabouts of the abductee in the manner of humans tagging animals in the wild to follow their wanderings. Corroborating evidence of this purpose is the occasional assurance of the aliens to the abductee that "We will know how to find you." In the words of the pastor:

> There is a separate wand which wields the needle which they stick in your nostril. Having thought about it, I'd venture to express that I have seen a little ball-thing that they install, remove or re-install. It's ceramic-like only more toward slivery-white. It's quite slender, and some inches in length. The crunching sound is constant whether they are sticking it into you or taking it out. Sometimes the placement seems to affect the sight of the eye that they've placed the little ball behind.

* * *

Should any of this seem too weird to believe, I reprint an episode detailed in a previous volume of mine, *Our Interplanetary Future*. It is an episode first published by Paul R. Hill in his book. *Unconventional Flying Objects*[238] Hill was a respected scientist, first with the National Advisory

Committee for Aeronautics (NACA), then with the National Aeronautics and Space Administration (NASA), when it superseded NACA in 1958. This led to his ultimate construction of the first flying platform that could be, and was, flight tested. He had personally witnessed two UFOs, one a USO (submerged instead of flying); it was spinning at a rate that he calculated would have subjected passengers to over 100 g forces, nothing that any Earthling could survive). He soon realized from his maneuvers that the fine controllability of his platform might, on the same principles, control maneuvers of "unconventional" flying objects, a term he preferred to "unidentified".

The official policy of both NACA and NASA was that UFOs did not exist, and Hill's name could not be used in connection with his sighting or in any way that could implicate NACA or NASA. He was very well respected, and obviously, from his writings and rigorous analyses has the mind of a scientist, one not easily fooled by eyewitness or any other kind of report.

The following does not involve Nordics, and is included only as an example of how strange the behavior of aliens might be, and as something to consider by anyone inclined to dismiss the above summaries out of hand. Any reader may find them unbelievable, but, in my opinion, should not do so without considering other reports such as this, which has passed muster with a highly disciplined scientific mind.

DeWane Donathan, a young married man living in Hartford, Indiana said he never believed in such things as UFOs until it happened to him and his wife, which it did one evening in late October, 1973. As they were headed toward their home they saw what first appeared to be a reflection from a tractor. But as they got closer, they could make out two figures that looked like they were dancing to music in a circle in the middle of the road. "It didn't look like they wanted to get very far apart from each other," Donathan told a reporter, "When they turned around and looked at our car, they acted like they couldn't walk off the road. They looked like they were skipping, but didn't have their feet in front of them... they had their arms in front of them." According to Mrs. Donathan, who was driving, the creatures were of slight build, straight of form and about 4 ft. tall. She said their feet had boxes on them, a little larger than a shoe.

Business man Gary Flatter, who happened to be in the police station when the Donathans' call came in rode with the sheriff to the area they described, but saw nothing. They returned in two vehicles. As Donathan approached an area south of the scene described, he was delayed by a parade of small animals crossing the road. Donathan counted 6 or 7 rabbits, a possum, a raccoon, and several cats. Flatter looked around and spotted the two creatures, silver suited, standing in a plowed field in the approximate area from which the animals had come. Whether Flatter, or anyone else, thought there was a connection between the animals and the two humanoids, we have no clue. We can draw our own conclusions, or none. In any event we should be grateful to Flatter for his detailed report.

They were, he says, 75 feet away and facing him. He estimated their height at 4 feet, and they were dressed in tight-fitting silvery suits that glared in the light. Illuminating them further with his spotlight, he was almost blinded by the reflection. He says the creatures didn't like it either, as with a hop they turned their backs to the beam. He turned it off. He then favors us with this highly interesting description of the creatures and their bewildering behavior.

> Their heads were egg shaped and fitted with what looked like gas masks with a garden-sized hose connection to the lower chest. The feet were "square with the heel a little over the back," having the approximate dimensions of 6x3x2 inches. As they continued to hop, they moved up and down in slow motion. They seemed to use no muscular effort when they jumped, but moved as though skipping rope. Flatter says the motive power seemed to be in the feet. "They would move up about three feet off the ground, then go back down all in slow motions… they might move an arm, but not much." The fourth time up, they simply flew off! "They flew like a helicopter in feet down position. They just flew off into the darkness and I couldn't find them with my spotlight. I did see some red trace-like streaks coming down, and that was all.

The following day seven imprints were found in the hard ground near the Donathan sighting that seemed to have been made with a nearly square heel. They were 3 inches wide and an inch deep. The feet of the investigators left no imprints.

Flatter has been described as a man who would not joke about something so serious. When Donathan was told of Flatter's sighting of the two creatures, he said that he was relieved that he and his wife were not the only ones to have seen them.

Is this one more bit of evidence that the space visitors have indeed been able to nullify gravitational effects? The parade of animals appears to fit in well with numerous descriptions of the aliens as collectors, collection being part of their role as investigators of Earth and everything about it.

I mention one more. It goes back quite a ways. It was reported in the *Lansing State Republican* newspaper in April 1897 that on the morning of the 17th of that month, in Williamston, Michigan, almost a dozen farmers saw an object maneuver in the sky for about an hour. It then landed and the pilot, almost 3 meters tall (about 9'9"), practically naked and suffering from the heat, emerged. The report stated that *"His talk, while musical, seemed to be a repetition of bellowings"* One farmer, upon going near him, was struck a blow that broke his hip.[239]

CHAPTER XXVI

TIMOTHY GOOD AND THE CASE OF ALBERT COE

WE COME TO A RESEARCHER, Timothy Good, who must rank very high on anybody's list of the best. One of his books most germane to our subject is *Alien Base*. The base is Earth. He rarely uses the word "Nordic," because, for one thing, he is usually quoting witnesses who know nothing of the use of that term to describe extraterrestrials. Sometimes, he, tellingly, uses it anyway as a descriptive term.

The episodes summarized here are all from *Alien Base*. Before discussing the witness who is the subject of this chapter, an explanation is in order.[240]

I try to use here and elsewhere only episodes that seem the most credible, that make the best case for believability. But I will leave that for the next chapter. I cannot necessarily say that for this first witness. One of the things he has to say however is so close to the target for our focus here I feel that even if it is not true, the story is very detailed, and the mere fact that someone, in his narrative, has articulated the very question we deal with, and given much thought to it, makes it worth repeating. If the narrator is giving us facts, the relevance is self-evident. If not, I feel it worth knowing that the problem has at least occurred to someone else who has invested time and thought to the subject.

Good's informant was Albert Coe, whose story in itself is quite exotic. But it is he who relates to Good the story of the extraterrestrial, whose name was given as Zret. Coe was, in 1920, when he says he first met Zret, sixteen years if age. Though Good does not tell us exactly when Coe's interviews with Zret first took place, he states: "Albert Coe's meetings with Zret—and others of his group—continued into the late 70s at the rate of 10-12 times per year."[241]

Good leads into the episode with a few paragraphs of explanation concerning rumors that persisted for a few hundred years about a group of rather odd folks, some of whom were seen in the late 19th Century emerging from time to time from the forests in the vicinity of Mount Shasta, California. Their mission was often to trade gold nuggets for basic commodities. The nuggets were worth far more than the items they bought, but they would accept no change, explaining that the gold was of no use to them. An author, named Wishar Cervé, wrote about these people in 1980 in his book entitled *Lemuria: the Lost Continent of the Pacific*. It is presumably this book on which Good relies for this introduction. That introduction is in itself germane to our subject, and whatever doubts may arise from the narrative of any of his informers, I see no reason to doubt the authenticity of Good's account of what was told to him.

Good writes that these strange individuals were described by local townspeople as "tall, graceful, and agile, with distinctive features such as large foreheads and long curly hair." They were also said to wear unusual clothes, including headdresses with a special decoration that came down from the forehead to the bridge of the nose. Good also wrote:

> On some occasions, powerful illuminations were observed in the forests, and strangely beautiful music could be heard. Invariably, when an investigator approached the area, he would be met by a 'heavily covered and concealed person of a large size who would lift him up and turn him away' from the

area. Other intruders reportedly were affected by some invisible influence, causing them to become temporarily paralyzed.[242]

The narrator, Albert Coe, claimed he met Zret by helping him, with his companion named "Rod," as Zret lay injured in a mountainous area near Ontario, California. Coe noticed some very strange things about Zret at once. Later the injured man claimed, according to Coe, to have come from a planet named Norca, revolving 85 million miles around Tau Ceti, about 11 light years from our solar system. In cosmic terms that distance is very short, though each light year is almost 6 trillion miles. Coe at first thought that Zret was having hallucinations.

He claimed to have been in a plane that crashed nearby, but Coe realized there was no place to land, however he was astounded to see a disc, not a plane, about 70 or 80 feet wide. Coe said that the injured man accepted his aid only on condition that he would not divulge what had taken place that day, even to Rod who had returned to the tent he and Coe shared. The reason was to be explained later.

There are many other details related by Good involving that first meeting, and the later communications about a second one. Though they seem to add to the credibility of the tale, we skip to the story of Zret as allegedly given to Coe.

Zret explained, according to Coe, that about 14,000 years ago the planet, Norca, began dehydrating slowly, and that drastic action was necessary to save the race. The only solution was to find another solar system. There was an exploratory mission to Earth, and contact briefly established with Cro-Magnon humans. It was decided that the Norcans would colonize Earth.

According to Zret, says Coe, 243, 000 Norcans left their planet in 62 huge spacecraft. Due to some misfortune, or miscalculation, all but one were drawn into our sun. The one crash landed on the planet Mars killing many, but 3,700 out of the 5,000 or so passengers survived. Their stay on Mars lasted about 900 years, following which 'succeeding generations,' launched probes to Venus and Earth. Bases on Venus were established to study its peculiar atmosphere. But the main colonization was concentrated on Earth at a number of places, including, in chronological order: the mythological continent of Atlantis; the Cuzco Valley in the Andes; the legendary continent of Lemuria, at a point about 1000 miles east of the Marshall Islands, northern Tibet, and finally, Lebanon.

There then follow in the summary by Good of Coe's recounting of the narrative by Zret: *Norcans reproduced with native inhabitants. Irrespective of skin pigmentation, Zret explained, the indigenous Earth people at that time had black or brown hair and eyes, and the interbreeding led to a blond-haired, fair skinned people.*

There are a number of things in this narrative that will bear some discussion. The skin, hair and eye color, from our perspective may be the most interesting. But we start with a focus on the Cro-Magnons, the name allegedly used by Zret.

The first skeleton of a modern type human was discovered by a French geologist in March 1868 in a rock shelter named *Abri de Cro-Magnon* in Les Eyzies, Dordogne in southwestern France. Four fossil adult skeletons, one infant, and some fragmentary bones were excavated and an occupation floor was exposed. Certain details of the burial led scientists to believe that the burials were intentional as were the placement of pendants or necklaces on the bodies. The burials were dated to about 30,000 BP.

Similar specimens were subsequently discovered in other parts of Europe and neighboring areas. The term "Cro-Magnon" soon came to be used in a general sense to describe the oldest modern people in Europe. By the 1970s the term was used for any early anatomically modern human wherever found, even to the cases of remains in Israel and various Paleo-Indians in the Americas.[243]

Caves and other sites containing art or artifacts, obviously of more modern humans, were found in southwestern France, and before about 1980 their "trail" had been pieced together. Archaeologists were bewildered about their place of origin, which was obviously not in Europe. It was, as one said, "in-dubitably non-European."[244]

It was only later that more current data[245] concerning the migrations of early humans has contributed to a refined definition of the term "Cro-Magnon." It remains an important term within the archaeological community only as an identifier for fossil remains of similar type in Europe and adjacent areas.[246]

The important point in the above is that Coe's interviews, he says, lasted through the 1970s. At least through the 1970s, the term Cro-Magnon was used for any modern humans, no matter where located. It meant something different in the 1970s than it does today.

We will need to look at Zret's claim, real or imagined by Coe, concerning coloration of skin, eyes, and hair, a subject for which we will save the next chapter. But first let us look at some rather "outrageous" claims of Zret, the ones concerning his clan using Venus and Mars as a base for research and see if they are as outrageous as some may think on first blush.

"Today," said Zret, our basic home is the highland of Venus, although a good part of our research is still conducted on Mars, especially electronic probes, for its thin atmosphere and peculiarity of magnetic fields lends itself as an ideal laboratory."

Zret claims that upon leaving Mars his clan's probing of Earth and Venus resulted in their choice of Earth as the main place of colonization. If so it was an almost inevitable, choice. Earth and Venus are similar in many ways, such as size, gravitational strength, and time of journey around the Sun. But there are many differences. Venus has an atmosphere, but it is almost all carbon dioxide, and very thick it is. The atmospheric pressure is almost a hundred times that of our Earth. Further, the temperature is about 740° Fahrenheit, a searing heat existing on both sides of the planet. The reason for the uniformity is a very dense cloud cover that prevents radiation of heat. How could the Norcans possibly have survived on Venus? Almost certainly, their native planet had no such atmospheric pressure or temperature.

Presumably they could do so the same way our own scientists plan on building a work and exploratory station on the Moon. They have written at length about such a Moon base. I, and I believe almost everyone else, has seen artists' renderings of this base of the future. The Moon's temperature hits a high during the 15 Earth days of sunlight of about 260° F and it is not steady. The temperature during the 15 or so days of darkness is—280° F.

If our scientists, after a mere 500 years or so, feel they can create within a large dome of the proper material, enough protection from the outside world in which to live and work on the Moon, why cannot a civilization several hundred thousand or more years ahead of ours handle a temperature of 740°? As for the atmospheric pressure, the Moon is almost a vacuum. Building up an atmosphere within the bubble on the Moon is probably more of a challenge than reducing the atmosphere within a similar bubble on Venus.

The late Carl Sagan, scientist and author, specializing in planetary sciences, went further. He claimed that use of appropriately grown algae "would in time convert the presently extremely hostile environment of Venus into one much more pleasant for human beings."[247]

Mars would seem to have been a lesser challenge than either Venus or the Moon. The daytime temperature averages 50° F, but can each 70° or 80° F; not at all unpleasant, but the nights are often—170° F. The comparison with the Moon should be sufficient to have us understand that a civilization hundreds of thousands of years ahead of us could well have the technology to establish bases on Mars or Venus, for research.

Coe says he told no one about the meetings until 1958, when with Zret's blessing he told his wife. She at first thought he was kidding. She wanted to meet Zret, but Coe said that was out, "You see these people are very serious", he explained to Good, "They have very good reasons to be, and I wouldn't want to be the one to let the secret out." If the request was truly answered so abruptly and definitively, I could find it suspiciously defensive. It could, to me, make the entire story more questionable than the talk of Mars or Venus. If Zret had already told Coe that he could, with Zret's blessing, tell his wife of the relationship, why could he not have made a simple inquiry of Zret about a meeting with her? But that is my subjective reaction, and perhaps insufficient upon which to base a judgment on the whole complex ball of wax.

I would not be shocked to learn that there was some truth to this tale. Neither would I be shocked to learn that Zret was Coe's "Harvey," an imaginary six foot tall bunny. The overgrown rabbit was realistic only to his human companion through life, in the stage play named Harvey. It ran on Broadway in the late 1940s. The human was named Elwood. Was Coe a real life Elwood?

In 1977 he had a session with a psychiatrist, a Dr. Berthold Schwartz. How it came about Good does not tell us, but Good did listen to the 90 minute tape of the meeting. Coe told the doctor that he had not suffered from delusions, encephalitis, hallucinations, or paranoia, nor had he spent time in a mental hospital. Good also read Coe's book. *The Shocking Truth*, self-published in 1969, and used both that book and the doctor's transcript as references.

Coe's summaries of Zret's narration to him is very detailed, containing much that strains credulity, perhaps nothing more so than the wanderings of the Norcans. Yet, given the proven facts of UFOs and extraterrestrials, the work of our own scientists, and the obviously great edge these other beings have in science and technology, very little if anything in his tale seems out of the realm of possibility. My own negative reaction to Coe's instant and peremptory negative reaction to his wife's very understandable request is a purely subjective matter, and I still cannot dismiss the entire matter on the facts at hand. If the story is contrived, Coe has done a tremendous amount of study and thinking to pull it off, but that would, relatively speaking, be a small cause for wonder.

Timothy Good is not naïve. He goes where credible facts take him. He is not afraid or even hesitant to state openly his belief in things that make many, scientists and persons of renown, not to mention the U.S. Government, scoff. Nor does he hesitate to reject matters that many of the Ufology community accept, nor to state his indecision where he is in fact undecided. What does he say about this one?

His conclusion: "I conclude that he told the truth; at least, the truth as he believed it."

Which tells us nothing, but, I must confess. I'm sure I could be no more certain than Good, even if I had his closer knowledge of available facts. The fact that Coe's story is not impossible does not mean that it is true. And it is interesting that he addressed the question of skin coloration, no matter how sketchily. In his narrative, the words come from his extraterrestrial friend, real or imagined, not from himself. I do not think I could, nor should anyone else, pass judgment until we look at the scientific facts of color change of hair and eyes, as well as skin, about which we have already dwelt at some length.

CHAPTER XXVII

BLUE EYES AND BLOND HAIR

So let us turn now to the subject of coloration and the changes resulting from reproduction with native inhabitants. Recall that Zret is reported to have said *Norcans reproduced with native inhabitants. Irrespective of skin pigmentation... the indigenous Earth people at that time had black or brown hair and eyes, and the interbreeding led to a blond-haired, fair skinned people.*

The phrase "irrespective of skin pigmentation" is of special interest. It may indicate that 14,000 years ago some of the Earthlings had lighter skin than others, light brown or tan, perhaps. It is indeed possible that the skin might have been changing and that the hair and eyes, under the control of different genes, had not, or that the changes, in any event, were not in tandem. Zret does not specifically mention change in eye color, but in mentioning the color brown as that of the Earth inhabitants, the inference is there, and I will assume he was including eye color change. If he wasn't it makes some, but little difference.

It would be entirely possible that, as Coe quoted from Zret, the change in hair or eye color came later. Eye colors other than brown have been found to only exist among individuals of European descent.

African and Asian populations are typically brown-eyed. In 2009 a team of researchers studying the gene known as OCA2 published results demonstrating that the allele associated with blue eyes occurred only within the last 6,000-10,000 years within the European population. To date, eight genes have been identified which impact eye color.[248]

The subject of hair color is more complex. Blond hair made a sudden appearance 11,000 years ago, in two parts of the world and is controlled by two different sets of genes. It has caused almost as great a mystery and has invited almost as much speculation as the great transition, the appearance of pale skin in Europe, and the appearance of blue eyes.

Black is the color of most of the people on Earth. Blond, brown, and red hair, until recent years, were believed to be found only in Europe and among those of European ancestry. According to a study published in 2012 hair color is largely determined by the amount of two pigments; brown-black eumelanin, which predominates in black and brown hair, and red-yellow pheomelanin in red hair. Blond hair contains low levels of both pigments. There is still much uncertainty about the genetics of human hair color, but some researchers believe that there are at least two genes that determine whether a person of European descent will have brown, blond or red hair.[249]

But blond hair, it has been found, has not at all been restricted to Europe. Indigenous people of the Solomon Islands in the South Pacific, who have some of the darkest skin outside of Africa, unlike most other tropical populations, also have a high prevalence of blond hair. With up to 10 per cent of the population being fair haired, they have the highest proportion outside of Europe. And contrary to what was first thought, it is not attributable to European explorers and traders in preceding centuries. The underlying genetics have been found to be different. According to a study published in 2012, among others, it is truly a case of convergent evolution.[250]

Apart from blond hair in the southwest Pacific, there are varying numbers of it in most other places in the world, including North and South America, Australia, New Zealand, and South Africa, but that is generally attributed to migration from Europe from the 16th through the 20th Centuries.

Blond hair occurs most frequently in Europe. The Solomon Islands and Melanesia have different genetics that are found to underlie it. The mutation that caused the blond hair in the natives of the Solomon Islanders is believed by researchers to be the same that caused it in Fiji and other areas of the South pacific, and is estimated to have occurred about 10,000 years ago.

Research carried out by three Japanese Universities date the mutation that resulted in blond hair in Europe at about 11,000 years ago, during the end of the last ice age.[251] Peter Frost finds the same approximate date, claiming further that the appearance of blue eyes and blond hair in some northern European women "made them stand out from their rivals at a time of fierce competition for scarce males."[252] Still another researcher says it first appeared in Lithuania about 3000 BC (5,000 years ago), and through sexual selection spread quickly through Scandinavia, while another claims that light hair was already in southern Siberia during the Bronze Age, or about 5500 years ago.

It has also been found that blond hair is most frequently found among populations of northern Europe; that pigment of hair and eyes is lightest around the Baltic Sea and that darkness of both increases regularly, almost concentrically around that region.[253] As shown in illustrations 46 and 47, the distributions are similar, but not at all identical

Though some claim that alleles for hair or eye color influence skin color, skin color has more often been found to be influenced very weakly by those various alleles apart from the ones for red hair or blue eyes. Some have no effect at all on skin pigmentation.[254]

Others claim that the color diversity resulted from random factors including genetic drift, meaning the random mutations of genes spreading through marital combinations unrelated to any other factors; founder effects, meaning the separation of a few light haired people from the others for any reason and consequently spawning a large population of mostly light haired people; and relaxation of natural selection, meaning no survival benefit or advantage attaching to light hair, etc. But, says Frost, such factors "could not have produced such a wide variety of hair and eye hues in the 35,000 years that modern humans have inhabited Europe." The hair-color gene, he states, has at least 7 alleles that exist only in Europe and the same is probably true for the eye-color gene,[255] and that If we take the hypothesis of a relaxation of selection, nearly a million years would be needed to accumulate this amount of diversity.[256]

Moreover, Frost continues, "it is odd that the same sort of diversification has occurred at two different genes whose only point in common is to color a facial feature"[257]

So evolutionary forces affecting both hair and eyes must have been of a non-random nature. But, asks Frost, why? And how? The geneticist Luigi L. Cavalli-Sforza may be among the most scholarly, and prestigious to advance the notion of sexual selection at work here. This occurs when one sex substantially outnumbers the other among individuals ready to mate. The sex in excess supply has to compete for a mate. Darwin's example of the peacock's tail has become the most publicly popular of examples to represent Darwin's explanation. The peacock grows a beautiful, colorful, highly visible tail of feathers because it attracts females. It serves no other purpose, but those males whose genes are perpetuated are those with the fanciest tails, and the size and beauty increase accordingly, it being those who outdo their peers in that department whose genes live to compete with the next generation.

How does that fit in to our conundrum here? In northern Europe in prehistoric times, says Calli-Sforza, the males for various reasons, including the dangers of hunting ferocious animals, were outnumbered by females. Why did they not practice polygamy? Because it was too expensive, requiring too much meat and too many victories over the animals. Why did blonds seem to win more of the competitions for a relatively scarce male? Because being initially standouts among so many dark haired women, they were more attractive to the men.

That goes only so far, says Calli-Sforza, and when blonds equal or surpass the dark haired beauties in number, there is no longer any advantage. Experiments with modern humans, namely men being asked to judge pictures of women and to state their preference for marriage, will increasingly pick the dark haired ones the scarcer they are among the more numerous blonds in the stacked deck of pictures.

In scholarly language it sounds more convincing, but has provoked much comment, over 600 on one web site. Some of the comments have a modicum of objectivity and reason, most are highly emotional, and more than a few are overtly racists, some odiously so.

231

We need not get into the merits of it. But we need to ask a few questions. According to Frost "it is odd that the same sort of diversification has evolved at two different genes whose only point in common is to color a facial feature." The suggested answer may, or may not explain the spread of blond hair after the initial appearance, meaning after the mutation that caused it. With due respect, the question Frost asked is not the question he answered. If the inhabitants of northern Europe did not interbreed with others who already had blond hair, and if there was no survival advantage to blond hair, which the scientists say there is not, why and how did the mutation come about in the first place.

If we were talking only about one mutation, and one facial feature, it could well have been a mere happenstance, a random occurrence such as happens so often. But Frost himself says "it is odd that the same sort of diversification has evolved at two different genes whose only point in common is to color a facial feature." He also states that if the sexual selection hypothesis is true, the European-specific alleles, meaning those for both blue eyes and blond hair," would have arisen almost entirely during the last ice age (25,000-10,000 BP)."

One occurrence could be random. With two it looks a lot less likely. But there are more than two. As we have seen earlier, and in this chapter, neither allele has any bearing or connection with the genes affecting skin color. Nor do they have any effect on each other. It seems, according to many researchers, that much of the skin color change occurred about 11,000 BP. Others say between 50 and 20,000 BP. We have seen some evidence from LA Marche that about 11,000 years ago the inhabitants of central France were quite possibly still exhibiting characteristics of both Caucasians and Negroids. So we have three different genes or mutations affecting three different aspects of appearance among northern Europeans and at least one, for hair color, affecting southwest Pacific natives, in ways similar to one in Europe. Could all three of the European mutations have been random? Is it a quartet of coincidences that all such changes occurred randomly within the last ice age? Or would it do less violence to reason to assume that one cause underlay all three, such as aliens from space, seen by prehistoric humans as gods?

H.R. Ellis Davidson refers us to the blond haired man Jarl from the poem Rigsbula in the *Poetic Edda* from Medieval northern Europe.

Jarl is considered to be the ancestor of the dominant warrior class. Also in Northern European folklore, "supernatural beings value blonde hair in humans. Blonde babies are more likely to be stolen and replaced with changelings, and young blonde women are more likely to be lured away to the land of the beings."[258] The beings. We never seem to know what to call them. It may all be significant; or it may not. Your choice.

The following two illustrations, 46 and 47, may assist in understanding the geographical relationship to blond hair and light eye color, respectively, in the area of northern Europe.

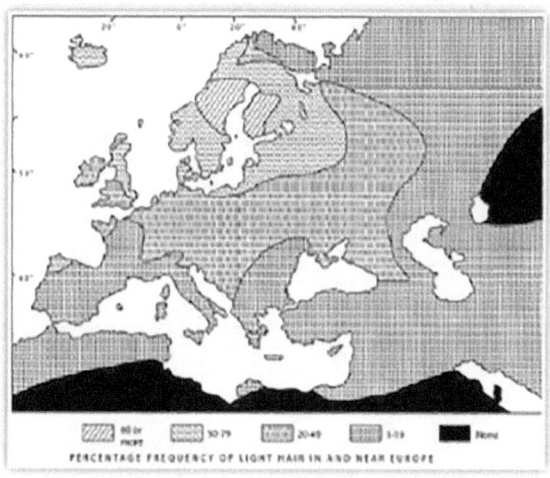

Illustration 46: Credit: Hair-color diversity in and near Europe (after Beals & Hoijer, 1965, p. 214. Beals et al., Further credit: "An Introduction to Anthropology," 3rd ed. Published by Allyn and Bacon, Boston, MA. And Pearson Education.)

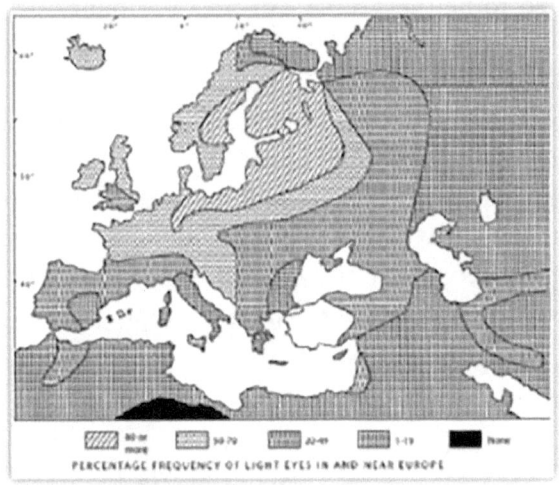

Illustration 47: Credit: Eye-color diversity in and near Europe (after Beals & Hoijer, 1965. p. 213: Lisa Raffensperger http://www.newscientist.com/article/dn21779-blonde-hair-evolved-inde-pendently-in-pacific-islands.htmlettal.): "An Introduction to Anthropology." 3rd ed. Published by Allyn and Bacon, Boston, MA. Further credit: Pearson Education.

What it all adds up to is that Coe's repetition of Zret's narrative, assuming Zret exists, is not at all at odds with the scientific investigations. The timeline, in short, is not all askew and can be safely fitted into the scientific findings. Does it prove anything? Not necessarily, but we are left with several choices: If it was contrived, Coe made a very lucky guess about the anthropological history; or he did a lot of research—most unlikely as much of the above findings were not even dreamed of before he told his tale to Good and others. Good's book was published in 1998. Or third, he is accurately repeating what Zret told him. Again, take your pick.

CHAPTER XXVIII

OTHER CASES OF TIMOTHY GOOD

WE TURN TO OTHER EPISODES and descriptions of Nordics or of those who so appear from the investigations of Good. They are culled from many other cases he summarizes. In choosing which of his many cases to include here. I have not used as criteria the degree of interest that they may arouse, nor the educational value in the field of Ufology generally, but only those cases containing descriptions of Nordics.

It is not always a clear cut choice. Good, who is thoroughly steeped in the subject, refers to the "bewildering profusion of alien species" described in his *Alien Base*. That volume is only one of six he has authored, and there are other authors as prodigious as he. Good estimates that at least a dozen different species have been here at one time or another, though not all may have come from different planets. But maybe they did. They certainly did not all come at different times from each other, though how much overlap there was, or is, no one can safely say. Neither do we know how long each of them has been here. Some are clearly different from the Nordics, the major interest for us, but sometimes it is a close call. Some of the creatures that have been here can best be described as grotesque by our standards, but there is nothing grotesque, by our standards, about the Nordics. They apparently do indeed look like us.

The circumstances of the sightings, meetings or confrontations will be kept to a minimum, but will hopefully be of sufficient provenience to paint a somewhat vivid picture.

One day in May 1940 near the village of Townsend, southeast of Helena, Montana, Udo Wartena, then 37 years of age, was doing work at his goldmining claim. He was astonished to suddenly hear and see a strange craft hovering a short distance away. A circular stairway came down and a man descended from it. Wartena went to meet him, but the stranger stopped about ten feet away. At his request, Wartena gave the man permission to take water from the stream running nearby, and he was invited to enter the craft. The two entered a room in the hovering craft where a second occupant was present. As Wartena put it:

> There was an older man already in the room, plainly dressed, but his hair was snow-white… the younger man's hair was also white. Both men were extremely good-looking, with perfect, almost translucent skin, and appeared to be very youthful and strong.

Wartena explained to Good what he had been told of the craft's propulsion system, but Good confessed that he did not fully understand the explanation. "Perhaps the most impressive aspect of the encounter for the witness was the strong bond of friendship he felt towards the cosmonauts. They were men, he stressed, 'just like us and very nice chaps', and felt even 'love, or comfort' in their presence"[259]

While serving with the U.S. Army in 1946, Allen Edwards, an academically trained portrait painter, was living with his wife in Petersburg, Virginia. Shortly before he was due to be discharged, he was admitted to the Army hospital at Camp Lee, Virginia for treatment of a minor ailment. He was given a private room, and the next morning went down the hall seeking someone to chat with. A number of patients were siting by their beds, but one stood out from the others. Even from 50 feet away, said Edwards, there was something unusual about this man. Edwards seated himself on a vacant bed where, trying not to be too obvious, he could study "this amazing man." The description given by Edwards to Good, effusive though it was, was similar to many other descriptions of Nordic extraterrestrials, something, this stranger never claimed to be.

> Never in my life had I seen such beauty, yet there was absolutely nothing feminine about his features. They were perfectly formed. His forehead was extremely high, the fine veins showing faintly through the transparency of the skin at the temples. His blond hair seemed to glow with an inner light of its own; in fact, his entire head seemed to be radiant, whether from the beauty of his complexion or some mysterious factor I do not know. His eyes were softly blue beneath the pure whiteness of his brow and seemed to me, to be filled with great compassion. His nose was perfectly shaped and the coloring of his cheeks had a freshness and purity that I had never beheld in any human being.
>
> The extraordinary height of his forehead amazed me but the physical characteristic that I found even more unusual was the depth of his head from the forehead to the back. This was definitely an abnormality according to all rules of skeletal structure and yet, as I continued to stare, I realized that for the first time I was seeing perfection.

Six weeks later, as Edwards was walking along the main street of Petersburg, Virginia, he saw the man walking in the same direction on the other side of the street. The conversation itself, interesting though it was, has no bearing on our particular interest in him. Of most interest is Edwards' initial impression. He told Good: "I found myself looking up at him: I am fairly tall but he towered above me."

Most, but not all of the extraterrestrial Nordics, are unusually tall, and such height is one more attribute that strengthens the case for it.

There were other meetings, which led to other experiences that probably convinced Edwards that the man was an extraterrestrial, though he was never so unequivocal. He said that he could not claim to have been in contact with beings from outer space, as the people who impressed him to such a degree never identified themselves as such. "I have placed these people in the category of 'extraordinary' due to the unusual powers that they possessed." It is possible, says Edwards, that they are of this Earth, but have reached a higher state of development than the average.[260]

It was July 23, 1947, and José Higgins, a Brazilian photographer was surveying land northeast of Pitanga in the State of São Paulo, Brazil. He and the others were suddenly frightened by a large circular airship whistling as it approached from the sky. The other men fled. Higgins alone stayed. The craft landed about 150 feet from him. It was about 15 feet in height. He saw two strange looking people through a thick window. Soon a hatch on the underside of the craft opened and three "strange beings" emerged He described the beings:

> About seven feet tall, the beings were enclosed in a kind of inflated, transparent suit which enveloped them from head to foot. Through the covering, Higgins could see they wore shirts shorts and sandals. The clothing appeared to be made of brilliantly colored paper. They had large round, almost hairless heads, as well as large round browless eyes, and appeared identical to each other—like twins or brothers. The length of their legs was greater in proportion to their bodies than those of normal human beings.

They moved around him with amazing agility. He was invited aboard the ship, but after being given permission to get his wife to go along with him, he found a spot where he could observe them without being seen. What he saw was the strangers frolicking about like children, jumping in the air and throwing huge stones around. Higgins later told the press "I will never know if they were men or women... they were somehow beautiful and appeared in excellent health.

He said that had it not been for the workers who were there in the beginning, he might have thought it a strange fascinating dream.[261]

On April 11, 1952, a 24 year old French woman identified only as Rose C., was living with her father near the French town of Nimes. In the middle of the night she was awakened by the growling of her dogs. She went outside and was greeted by three very tall men of about 2.30 to 2.40 meters (7'5" to 7'10") in height and one normal sized man. The normal sized man spoke perfect French and acted a translator. To Rose, two of the tall men looked very athletic, the third, seemingly acting as their leader, was older than the others. The man of normal height said that the giants had come from a "faraway world." She was shown their craft hovering about three feet off the ground. She described it as enormous and shaped like a straw hat. She declined their invitation to go with them explaining that she had an elderly father and a young daughter to take care of. The normal sized man related to her how he had been contacted by these beings 20 years previously. Having no ties he accepted the invitation to live with them.

We are told nothing about the color of the skin, eyes or hair of these seven foot giants, but there is one postscript that bears repeating. The case was investigated by one, Charles Gourain, and published in France. He wrote that in his opinion, extraterrestrial encounters are always predetermined, and are not accidental. It is, he continued, "a genetic inheritance" which motivates the encounters between the extraterrestrial and us.[262]

In the late afternoon on October 21, 1954, Jessie Roestenburg was at home in Ranton, Staffordsire, England, with her little girl and her dog. Her husband was at work and her two boys were in school, though it was past time for them to have returned. She heard a loud hissing sound, loud enough that she thought it was a plane crashing. She went out to investigate and was surprised to see her two boys lying flat on the ground. One of them shouted "Mummy, there's a flying saucer." She replied, "Don't be silly," though she could see they were very frightened. She discovered that she "was not in control" of herself. She walked up to the water pump and turned around, but "it was as though someone else was making me do it. I wanted to look at the boys and ask them if they were alright, but I couldn't."

She then saw, suspended in the air a massive disk, shaped like a Mexican hat. She looked inside and saw two beings looking down at them. She continued:

> They were the most beautiful people I have ever seen, but they weren't human. Their foreheads were large in proportion to the rest of their faces, and they had long golden hair. I could only see them from the chest upwards, and they were wearing what looked like vivid blue polo-neck jumpers and what looked like fishbowls over their heads.

She recalled that though the figures looked like women, she felt sure they were men. They gazed down at her and the two boys with a seemingly stern, though compassionate expression. "I couldn't move. I was absolutely paralyzed. I wasn't frightened at that stage, but I was mesmerized... I felt all the tension go from me and I felt a sense of peace I have never felt since."

But her health began to deteriorate. She told her doctor she wanted to see a psychiatrist. The doctor told her that there was nothing wrong with her mind but that she needed to go to a hospital. She was found to be suffering from radiation sickness. Good met with Jessie on a number of occasions and is convinced of her honesty. He feels she still contains a sense of awe and wonder about the incident. In 1996 she told a news reporter,

> To this day I don't know what they were. I don't believe they wanted to do us any harm. They are far more intelligent than we are. We must have looked a pitiful sight, standing there next to a water pump while they were in a space ship.[263]

We go to Porto Alegre, Rio Grande do Sul, Brazil, on November 10, 1954. An agricultural engineer and his wife and daughter were taking a drive when they saw a landed disc. Two normal sized men alighted. They had long hair and were dressed in overall-like clothing. The two strangers approached the car, whereupon the wife and daughter fled the scene. As they looked back they saw the two men board the craft and take off.

A month later, on December 9th Olmira da Costa e Rosa was cultivating his crops at Linha Belha Vista, also in Rio Grande do Sul when he heard a noise like a sewing machine. The nearby animals panicked and fled. Olmira looked up, saw a strange man, and hovering further away an object shaped like an explorer's hat. There was one other man in the craft, and a third was examining a barbed wire fence. There followed somewhat friendly exchanges of gestures and smiles. Olmira was almost illiterate and knew nothing about "flying saucers." He believed them to be aviators from another country. To Coral Lorenzen, a famed UFO investigator, he described the men. As summarized by Lorenzen:

> They appeared to be of medium height, broad shouldered, with long blond hair which blew in the wind. With their extremely pale skin and slanted eyes they were not normal looking by Earth standards.

After the craft left, Costa e Rosa looked around the ground but found nothing. However, he said, for some time afterwards, the area where it had landed had the smell of burning coal.[264] The strangers had blond hair and pale skin, but we know nothing about their facial features or their eye color. Were they Nordics? It may be a close call, but being more of a lumper than a splitter. I vote yes. The pleasantries add something to the hair and skin. To me they "feel" like the Nordics I have read about elsewhere.

In July 1957, Professor João de Freitas Guimarães, a lawyer and Professor of Ancient Roman Law in the Catholic Faculty of Law at Santos, Brazil had a strange encounter in the proximity of nearby Fort Itaipu. Good has summarized the transcription of a TV interview of the Professor on Brazilian television on August 27, 1967.

According to Good's summary of the English translation, the Professor was seated on the beach about 7:15 P.M. when some sort of "high bellied craft, shaped like a hat headed for the beach, and as it arrived threw out a line to which were attached some spheres. A metallic stairway came down and two men stepped out and approached the professor. Both appeared to be completely human. The professor's description of the men:

> They were tall beings, over 1.80 meters (5'10") in height, with long fair hair extending to their shoulders. Their complexions were fair, they had eyebrows, and their appearance was youthful and they had light colored, wise, and understanding eyes. They wore greenish one-piece suits fitting closely at the neck, wrist and ankles.

The professor accepted their invitation to board the ship, and he was taken for a flight of about 30 to 40 minutes. He could not be more exact as his watch had stopped, a not unusual happening in such proximity of these craft. With some warning about contamination of our atmosphere with our atomic testing, the visitors departed.[265]

The case of Orlando Jorge Ferraudi is of unusual interest even in a field so filled with unusual reports and episodes. Suffice it to say that his entire story was investigated for eight years by four investigators working as a team.

They had him recount the story over and over, studied his account and his body language. They concluded that Ferraudi was totally truthful. Ferraudi says that on one night in August 1956 he was fishing on a deserted coast, as was his custom, near Buenos Aires, Argentina. About 11:30 P.M. he started to feel that someone was watching him, but when he turned his head,

> I saw him; a strange individual observing me. Comparing him with my 1.9 meters, I at once estimated he was more than two meters (6'6") tall. His skin was very white and he had very light-colored eyes, no beard or mustache, had short and neat hair, and he wore some kind of tight-fitting overall.

We, unfortunately, must leave him with that. If we are to keep our eyes on one of the most interesting tales of all in this field of Ufology, that of a Madame X, we cannot go running down rabbit tracks.[266]

At 2:A.M. on April 30, 1983, a woman identified only as Madame X was asleep in her bedroom in her home near Sospel, 20 kilometers northeast of Nice, France. She saw a red football-sized object, tried to awaken her sleeping husband beside her but was unsuccessful. When the light vanished, she assumed it was "ball lightning." She went to open the window in an adjoining room and upon returning to the bedroom was startled to see four quasi-human beings, about 1.75 meters (5'9") in height. She later described them to investigators who summarized her account:

> They were of a very athletic gait, with long pale faces and thin noses and mouths. The eyes were very elongated, with blue irises. Most surprising was the position of the pupils: rather than centered, they were close to the corner of the eye, giving them a 'cross-eyed' appearance. They had blond eyebrows. In spite of their peculiar features, Madame X thought they were handsome. Whereas all humans have a recess or indentation at the point where the nose is joined to the forehead, this was non-existent in them. Neither did they possess the little furrow which we all have between our nose and our mouth. They did have teeth.
>
> She did not see their hair, for their heads were covered by little skull-caps like those of frogmen, but with this difference: they were not part of the whole one piece suit, and merely covered the entire skull and ears... Their hands, gloveless were soft and delicate and a little larger than normal human hands.

Although Madame X said that she had not been afraid, she tried again to waken her husband, but was again unsuccessful. One of the men speaking in normal French told her it was useless. She invited the men to be seated. One of them asked, "Do you know who we are?" You're robots, she replied. The man smiled and held out his hand for her to feel. She noticed the skin was of soft texture, then said, "You are extraterrestrials."

The short meeting that followed was friendly. One aspect of it must be included. At their request, she followed them outside. They were joined by three other quasi-humans holding black spheres in their hands. She and the first three men watched as sepia-colored images about three meters (9'9" high) were projected onto the fog without any visible means of projection from the spheres or elsewhere. The film ran from prehistory to the Second World War. It retraced the wars, sometimes stopping for careful viewing. Madame X told them she was not interested in wars, to which they responded that wars were all humans knew about, and that they only knew this planet in that light. We might recall at this point the holograms describe by Steve Wilson and Jim Sparks

Some of the most frequently published photographs of flying saucers were taken by Apolinar Alberto Villa Jr. He was born in 1916, of Native American and mixed descent. He claimed to have had contact with extraterrestrials from the age of five. Though he failed to complete the 10th grade of school, he seemed well versed in mathematics, electricity, physics and mechanics. He also showed an unusual talent for detecting defects in engines and their parts. He worked as a mechanic, first in the Air Force and later for the Department of Water and Power in Los Angeles. At long Beach one day in 1953, he had a strong feeling to go down to the beach, something he did not understand. There, he says, he met a man about 7 feet tall. He was afraid and wanted to run away but the man called him by name and told Villa many things about himself. Villa then realized that this was a "spaceman" and of very superior intelligence. There was some talk about the nature of the mission of extraterrestrials to Earth, and many interesting facets of it were explained by him

On June 16, 1963, ten years after the previous meeting, some of Villa's space contacts told him telepathically to be at a certain time at a certain place. Villa complied and brought a camera with him; he planned on using the opportunity to photograph what he could. At 2:30 P.M. the craft landed on tripod legs:

Nine beautiful crew members-five men and four women- disembarked through a previously invisible door. These beings ranged in height from seven to nine feet, and were well proportioned, immaculately groomed and dressed in tight-fitting one-piece uniforms. The color of their hair ranged from 'fiery golden' to 'polished copper' to black.

Villa was told that they came from the constellation we call Coma Berenices, which, Good tells us, is notable for the large number of galaxies it contains. It is worth mentioning that, in addition to communicating telepathically, they spoke in both English and Spanish, Villa's native tongue. What they had to say was also often like what other Nordics, but hardly any other group of extraterrestrials, had to say. It was about peace and the necessity for it, and the futility of the destruction going on upon the Earth.

Some details after the meetings would also bear mentioning. Most researchers who spent time with Villa found him to be genuine. He never sought publicity, possibly, says Good, because of disturbing threats, one attempt on his life, and the helicopters that were often hovering over him. Good quotes Bill Sherwood who wrote "He certainly never tried to use his unusual personal experiences for personal gain. To me he always seemed humble and sincere, unimpressed by the attention he received from the Secretary-General of the United Nations, U Thant, who called him to discuss [for 40 minutes], his experiences with the extraterrestrials. Good's own conclusion: a growing conviction that "Paul Villa's story contains essential elements of truth."

Otherwise the details of his meetings, as well as the black hair of one of the visitors, we must leave without further comment. We are told nothing about skin color, and the black hair alone would not disqualify anyone from being a member of the Nordic group. The ancestry of some of them, we must recall, included the African.[267]

We end this summary of a few cases of contact reported by timothy Good with an episode involving one, Jim Evans, a former U.S. Air Force security officer at his ranch in Colorado. He and his family and others had experience with extraterrestrials beginning in 1975. On this particular occasion, and Good does not say when, though he says it was late at night, his description of the two beings involved is in accord with that of many others going back to about 1920:

> They were approximately 5'6" tall, and had on tight-fitting cloth had on tight fitting clothing, you know, like a flight suit. I noticed the clothing changed color, from brown to silver, but I don't know how. They were very fair, had large eyes and seemed perfectly normal, completely relaxed. They had blond hair with something over the head ... the thing that impressed me most was the eyes... their facial features were finer. They were almost delicately effeminate, completely self-assured.[268]

We are dealing with an extraordinarily complex situation. There may well have been interbreeding between the various alien species, which complicates even more the problems with categorization. Even apart from that probability, we have subjects who are blond haired, blue eyed, warm and friendly, but not tall. We have other cases of those who are warm tall and black haired. Sometimes we know about facial features, sometimes we do not. I suppose that anyone's choice as to who is, or is not a Nordic is as good as anyone else's.

I should like to close this brief summary with some words of Good about his own book:

> The bewildering profusion of alien species described in this book [which is one of six]—and more could have been included—suggests that at least a dozen different species have visited Earth. This does not necessarily mean that they all come from different planets. The diverse indigenous human and animal species abundant on Earth are an indication that other planets similarly support a variety of species.

There are scores of other cases reported in this one volume of Good's, but I was interested in those that involved at least a significant possibility of being of the Nordic aliens. If the Nordics did interbred with humans in the past, it is possible that other races did also. I have included only those cases which seemed to me the most credible to an average reader. In this I have relied only partially on my own instincts, but also on what I thought was not too strange or too difficult for a semi-skeptic to at least chew on. Perhaps, in any event, it all evens out.

There may be cases I have left out that were entirely, or substantially, true. There may be some I included that were substantially or totally false, though I sincerely doubt that. Please remember that the accounts I have included are much abbreviated, many details omitted. Before dismissing these cases as unbelievable, perhaps you can withhold judgment until reading the full account in the source material I have furnished in the bibliography or hear more on the subject, which there most certainly will be.

Meanwhile there is another set of stories we should consider. Not that I believe they are all entirely true. None of them can be entirely true, and some or all may be pure imagination. But maybe through all the barnacles they have picked up through the eons of their existence, at core there may be something solid, some seed, or fact upon which the sticky superstructure is built.

CHAPTER XXIX

RIPPLES FROM LONG AGO

We might start with the Bible. I have never before quoted the Bible, but as I have often heard, there is a first time for everything; a thousand pardons however for leading into it with the Greeks.

We call it mythology, but until the playwrights got hold of it, for the Greeks it was religion. The Greeks worshipped many gods. They believed their gods were superhuman beings and that they intermingled with humans. There was much love, or lust, exchanged and from such doings their heroes were born. The mythology is a collection of the religious stories of the ancient Greeks. Many of these beliefs have been preserved and passed down to us today, most especially through the great Greek tragedies. As was common in that time period in Europe and the Middle East. Judaism excepted, the concept of a god was usually that of a much more powerful form of human, but with all of the human faults and frailties.

Stories also abounded of heroes and fantastical beasts. The most courageous acts were from humans, or at least half humans. The concept of intermarriage or mating between gods and humans caused many demi-gods who would then continue to mate with humans themselves. Many parts of ancient Greece claimed some connection with godhood as did much of the world.

This mythology may be pure storytelling. But it may be rooted in real dramatic and erotic happenings in the past. There were good, even beautiful stories, but after many centuries, it is hard, if not impossible to tell fact from fiction. The Bible is more direct. As has been referred to in Chapter IV, the authors of the Old Testament had it right on the line. Genesis 6: 1-4 tells us exactly what happened—maybe.

> *When men began to increase in number on the earth and daughters were born to them, the sons of God saw that the daughters of men were beautiful, and they married any of them they chose. Then the Lord said, "My Spirit will not contend with man forever, for he is mortal; his days will be a hundred and twenty years." The Nephilim were on the earth in those days—and also afterward—when the sons of God went to the daughters of men and had children by them. They were the heroes of old, men of renown.*

Some things could still bear some explanation, or interpretation, of which there has been plenty.

What is a Nephilim? There is little need to get drawn into that controversy, religious or otherwise. The name "Nephilim" has various interpretations but is most often translated as "giants." As we saw in Chapter IV, Sitchin first interpreted it as "people who fell down upon the Earth from the heavens." He later acknowledged that in ancient Hebrew the word can mean giant. Hence we can take it as a given that "Nephilim" means giants. The union of the sons of God and the daughters of men produced a race of giants. It seems clear also that the term "sons of God" as used in Genesis would mean angels.

The New American Bible commentary suggests that the reference is implicitly to the paternity of Nephilim as heavenly beings that came to earth and had sexual intercourse with women. The footnotes of the Jerusalem Bible suggest that the Biblical authors intended the Nephilim to be an "anecdote of a superhuman race." It is noted that the Biblical passage implies that the Nephilim have inhabited the Earth in at least two different time periods—in antediluvian times "and afterward," namely both before and after the flood, something else the scholars have almost unanimously agreed upon.

If the Nephilim were supernatural beings themselves, or at least the progeny of supernatural beings, there is a theory that the "giants of Canaan" in *Book of Numbers* 13:33 were the direct descendants of the antediluvian Nephilim, or were fathered by the same supernatural parents. These giants were later said by the Hebrew observers to be as tall as Cedars.[269] Though there is ambiguity as to whether "the ancient warriors, the men of renown," means the "sons of God" or the offspring of the sons, most Biblical authorities seem convinced for various reasons that it is the offspring of the sons of God and mortal women that are referred to.

Whether they were angels or not I leave to others to thrash out. If they were anybody at all. I put my money on the probability that they were extraterrestrials of the type we call Nordics.

The next issue is the Lord's expressed intention to limit the life spans of the intruders/outsiders: *(Genesis 6:1-4)* "For behold, I will bring a flood of waters upon the earth to destroy all flesh in which is the breath of life under heaven. Everything that is on the earth shall die" though 120 years does not sound like a bad bargain.

Native global flood stories are documented as history or legend in almost every region on earth. Old world missionaries reported their amazement at finding remote tribes already possessing legends with tremendous similarities to the Bible's accounts of the worldwide flood. It has been estimated that altogether there are over 500 Flood legends worldwide. Ancient civilizations including China, Babylonia, Wales, Russia, India, America, Hawaii, Scandinavia, Sumatra, Peru, and Polynesia all have their own versions of a giant flood.[270] Though not so often mentioned is the very similar, but older version in Gilgamesh.

These tales are frequently linked by common elements that parallel the Biblical account. The overwhelming consistency among flood legends found in distant parts of the globe indicates they were derived from the same origin. The source has been stated often as the Biblical account. Little regard has been given the fact that when the glaciers, that covered much of the Earth melted at the end of the Ice Age, there must of necessity been widespread flooding. These glaciers covered much of Canada and the northern United States, for instance, with ice about two miles thick. Most sea levels now are about 200 to 400 feet higher than during the last glaciation period.

The date most often given for the end of the last period of glaciation is 12,500 BP. Unfortunately there is no way to scientifically link the flood of the Bible to the end of the last period of glaciation, or to anything else. The Biblical account seems obviously to make it much more recent, but who can be sure?

Again we need not delve into the specifics or any controversy. According to the authors of the Bible, the angels, if such they were, fell out of favor with the Lord and His answer was the flood. The story of Noah and the Ark is too well known to bear repetition here. Ultimately 10% of the angels/giants were allowed to stay and live in spirit, evil spirit, as a test for humans.

We turn now to another part of the world. In Harold Osborn's authoritative work, *South American Mythology*, there is the text of a narrative reported by Pedro de Cieza de León in the early 16th century.[271] We hear about some white men and of a god-like figure named Viracocha. We learn this by way of an interview with members of the tribe of Collao, long under the domination of the Inca. León, The Spanish explorer, heard Viracocha had by his alliance with the Inca become the master of many people of the Collao without fighting. Highland people all knew that the creator of all things was Viracocha and that his principal abode was in heaven.

León also wrote that "among the natives of the Collao are men of good understanding who show their intelligence when they are questioned. They keep account of time and know something of the movements of both the sun and the moon." According to León, many of the Collao also said that long ago in the Island of Titicaca, there were people with beards like "ourselves," referring to the beards of the Spanish explorers.

We also learn from León's work, created between 1532 and 1550, that long before the Inca had ever been heard of in the kingdoms, "these

Indians relate a thing more noteworthy than anything else that they say." After a period of hardship and much prayer to their gods, a light arose from the island of Titicaca, and all rejoiced. The narrative of León continues with this:

> After this had happened, they say that there suddenly appeared, coming from the south [the location of Lake Titicaca], a white man of large stature, and authoritative demeanor. This man had such great power that he changed the hills into valleys and from the valleys made great hills, causing streams to flow from the living stone. When they saw these things they called him Maker of all things created and Prince of all things, Father of the sun… This is the story that the Indians themselves told me and they heard it from their fathers who in their turn had it from the old songs which were handed down from very ancient times. They gave him the name Viracocha, which means "foam of the sea.[272]

"A white man of large stature." The early 16th Century was far too soon for influence either by the church or by Columbus and his crew. In the early 16th Century, and even more to the point, in earlier generations, what did these Native Americans know of white men? León wrote that according to the Indians, this man travelled the highland route to the north and that he worked miracles wherever he went. They told León that in many places,

> He gave men instructions how they should live, speaking to them with great love and kindness and admonishing them to be good and to do no damage or injury to one another, but to love one another and to show charity to all.

This may seem like boiler plate morality today, but judging by accounts of life in those faraway places in faraway times, it might have been something startlingly new. In most places, writes León, they name him Ticci Viracocha, but in the Province of Collao he is called Tuapaca [Thunupa] ... The Indians related that "after much time had passed they saw another man, like in appearance to the first, and that they have it from their forebears that wherever he passed he healed all that were sick and restored sight to the blind by words alone. Thus he was beloved by all." But working in a village called Cacha the populace rose up against him and threatened to stone him. They saw him raise his hands to the sky and the Indians "declare they saw fire in the sky which seemed all around them." The terrified Indians sought his forgiveness, and they "saw that the fire was extinguished on his command"

There is little in the narrative of León that sounds any more unlikely or impossible than eyewitness reports of modern people about UFOs and extraterrestrials. These include accounts of seemingly miraculous healing of serious injury, disease or debilitations: the expressions of concern by the extraterrestrials, especially the Nordics, for our seeming determination to engage in perpetual warfare; their admonitions to live in peace; the appearance of fire from exotic craft in the sky, and its disappearance; the emergence of UFOs from the sea, and their submergence under the sea.

The name used above, Thunupa/Viracocha, is a reflection of the uncertainty of the "mythological" structure of this geographical area. There was a later appearance of one who may, or may not have been the same person that we have just seen. The description we get about him comes from a different source, Antonio de la Calancha.

The description of Calancha as summarized by Osborne: "Thunupa appeared on the Altiplano in ancient times, coming from the north with five disciples. He was a man of august presence, blue-eyed, bearded, without headgear and wearing as cusma, a jerkin or sleeveless shirt reaching to his knees. He was sober, puritanical, and preached against drunkenness, polygamy and war."[273]

A man of august presence, and blue eyed. The balance of the description and the content of his admonitions are also suggestive of the modern Nordics. Admittedly, unlike the earlier appearance of Viracocha, he sounds like he could well be an apostle, self-appointed or otherwise.

The date of publication by Calancha is 1638, time enough after the discovery of the New World for the Fathers of the Church to have left their influence on the native populations, something unlikely, if not impossible in the case of the earlier description of Viracocha by León.

Still, the blue eyes, and the content of the speech are again suggestive. Much emphasis seems to be placed on the beard as evidence of European identity. Even León seems to agree. Nonetheless, it must be stated that there are busts of various prehistoric South American gods and heroes with beards.

The descriptions, in any event, are anomalous in that culture where the gods are depicted in stone sculptures and in verbal accounts as weird, frightening, and grotesque. Their behaviors are consistently described in ways we would call cruel. Viracocha alone sounds and apparently looked human. Illustration 48, below, is from Pokótia, south of Tiahuanaco. Osborn says that it is from the earlier of two periods of sculpture found in that area. Spanish chroniclers say the native population connected them with local creation mythology and belief in an earlier race of giants and a race of men turned into stone by the creator.

Illustration 48: Seated figure from Pokótia, south of Tiahuanaco
Credit: Charles Osborne, *South American Mythology*, 69.

One of a number of statues that stand before the small rectangular temples of roughly hewn stone slabs in the highland district Huila, Columbia is shown in illustration 49 below. According to Osborn, these are "often carved in the round, they are squat and powerful, square cut and formidable. They have round, staring and deep-set eyes, broad strong noses with wide nostrils and prominent feline fangs. They wear a loin cloth or a girdle and usually carry a mace or other object."[274]

We have no statue of Viracocha, by that or any other name. The descriptions we have of him however, obviously do not comport with the illustrations 48 or 49. We somehow cannot picture either of those gods doing what Viracocha is said to have done, or saying what he is reported to have said.

Illustration 49: one of the statues in the highland district of Huila, near San Augustin, Columbia.

Credit: Charles Osborn, *South American Mythology*, 108.

There is one further account from this culture germane our interest.[275] One, Juan de Santa Cruz Pachacuti-Yamqui Salcamayhua, repeating what he claimed was told him by the ancient Incas had this to say: Shortly after the beginning of the world, the age of darkness, the tribes from each of the four parts of the Inca Empire settled in various districts. This age lasted a "vast number of years."

There later came a shortage of land, and wars became frequent. It was an age of savagery, devils were present, and it was not safe to go out at night. The devils were ultimately driven from the land and sometime thereafter

> There arrived in these territories ... a bearded man of medium height and long hair dressed in a rather long cloak. They say he was past his prime, with grey hair and lean. He walked with a staff and addressed the natives with love, calling them his sons and daughters. As he travelled all the land he worked miracles. He healed the sick by his touch. He spoke every tongue even better than the natives.

They called him Thunupa Viracocharapacha, which means servant, or with other suffixes to Viracocha meaning preacher. But surely, said this Spanish informant, he was the glorious apostle St. Thomas.

Maybe he was; maybe he wasn't. Maybe he was a Nordic extraterrestrial, and maybe he wasn't. Maybe he was the figment of someone's imagination and never existed; or maybe he wasn't and really did exist.

Whatever the case, the most interesting part of that account is the claim that he spoke every tongue even better than the natives. ETs of many descriptions, not only Nordics, have left many Earthlings with the impression that they were addressed in their native language, English, Russian. Portuguese and Spanish most particularly.

Some of our people have been observant enough to note that the mouths of the ETs never moved, and realized that the visitors were very adept at mental telepathy. Apparently what is transmitted are thoughts, concepts, ideas, commands, and suggestions that the recipients automatically translate to their language. The visitors have shown themselves equally adept at reading the minds of our people, whether we speak or not. Many earthly humans have been unsettled to realize that. If the primitive, imaginative natives of ancient Peru are indulging in flights of fancy, then perhaps so are we, or at least our contactees. But we have much more basis for judging the credibility of our contemporaries, and some seem very credible indeed.

Progress in certain areas of mind control, far less advanced than the mental telepathy demonstrated by ETs, but relevant to it nonetheless, is worth considering by anyone inclined to dismiss such reports as pure imagination. Two tetraplegic participants, lacking, that is, the use of all four limbs, in May 2012 successfully performed reaching and grasping tasks using a robotic arm controlled directly through neural activity.[276] This was accomplished through BrainGate, a neural interface system to control two types of robotic arms developed at Brown University. It involves implantation of a device in the motor cortex part of the brain containing 96 electrodes. The electrodes record the neural activity associated with intended movement, and an external computer translates the pattern of impulses into commands that control the robotic device. We might recall, strange as it seems, that there was a time when we had to plug in our phones in order to use them.

The two participants used the arm to reach for and grasp foam targets that were placed in front of them. It was the first demonstration of neural control of a robotic device through three-dimensional space. The woman was able to pick up and drink from a bottle filled with coffee before successfully placing it back on the table, the first time she had been able to do so in almost 15 years. Of relevance to us here is the demonstration of the electrical nature of the brain impulses associated with thought, the ability to convey them to an external object built (or trained) to receive and understand the impulses and to be guided by them. This experiment is admittedly a far cry from mental telepathy, but it demonstrates the reality of the necessary ingredients for mental telepathy and a host of other innovations. It also renders very believable the ability of a technology thousands if not millions of years more advanced than ours to do what so many witnesses, over many years and perhaps many centuries, have reported.

Somewhat closer to mental telepathy is the research described in an article on page 32 in *Scientific American* for December 2012. The author,

David Pogue, described his experience, not with aliens, but home grown earthly scientists who are developing a "brain-computer interface." He writes: "Regardless of our native language or personal history, the same parts of our brain 'light up' when we think of certain nouns... The computer knows which brain areas are active for which qualities. The system can guess what number you're thinking of or which of 15 emotions you're feeling." What kind of electrical connection will we use in emulating the interpretive prowess of the computer? None, except as Pogue puts it: "the electrical device you were born with."

If a 16th Century Native American met a creature with thousands of years of technological advance, is it any wonder he thought the creature was a god?

The South American natives described above, include, it may be recalled, the existence of giants in their past. Almost all mythologies, if mythologies they are, include giants in the past. As an example we look at one more, closer to us in both time and space.

William Frederick "Buffalo Bill" Cody lived from 1845 to 1917. He was an American soldier, buffalo hunter, and showman. In his autobiography, he relates a tale about giants told him by Pawnee Indians. The Pawnee once brought into camp some very large bones, one of which the surgeon of the expedition pronounced to be the thigh bone of a human being. The Indians said the bones were those of a race of people who long ago had lived in that country. They said these people were three times the size of a man of the present day. He said further that they were so swift and strong that they could run by the side of a buffalo, and, taking the animal in one arm, could tear off a leg and eat it as they ran.

Cody says that these giants had six fingers, and that they denied the existence of the "Great Spirit." This so displeased the "Great Spirit" that he brought a deluge upon the land that submerged even the mountain tops, killing the giants.

Unfortunately we are told nothing of color of these strong men, or anything about marriage with native women. But the reference to these outsiders who were punished by the flood for their impiety is strikingly similar to the Biblical account.[277]

Such ancient texts and oral histories may mean nothing at all. Or they may tell us only something about the human capacity for imagination.

Or there may be a germ of truth or historical occurrences at their cores. Most scientists would, and do, say that such things prove nothing. Many psychiatrist and neurologists say that such tales can be explained as psychic phenomena of one sort or another. I am not a scientist or a neurologist or psychiatrist, but a retired lawyer, used to relying on common everyday experience to make judgments of credibility. The factors include often the number of witnesses, apparently credible, with no reason to falsify and no history of hallucination, who say essentially the same thing. I find the almost universal nature of sightings of giants, descriptions of "white men" where Caucasians would not be expected, and their preaching of things so similar to modern reports of the Nordic ETs, possibly all a bit too much for coincidence, and too much to be ignored.

CHAPTER XXX

MUSINGS OF A SKEPTIC

Almost everyone is familiar with the icon representing justice. It is the lady with the blindfold over her eyes and the scales, one in each hand. It is supposed to symbolize the impartiality of justice. It can be seen, if nowhere else, adorning thousands of courthouses around the land. It means she does not care who the parties to the controversy are. Their relative wealth, their heritage, their rank, their connections or anything else, except the truth or validity of their claim mean nothing. It is the merits of conflicting claims that are weighed on the scales without knowledge by the justice goddess as to the identity of the parties.

It is a beautiful concept, but laws are made by humans and they are administered by humans. Those who have had experience with the making or the administration of the laws often learn to their sorrow, others to their joy, that the reality may be something different.

The laws of nature are more predictable. People who fall from ladders will be injured according to the height of the fall, the strength of their bones, and other factors. But their names or ranks, their heritage or wealth will make no difference. The laws of gravity apply equally to one and all, without preference or prejudice. A rich person will fall just as hard as a poor one and a prominent person just as hard as a non-entity. A venomous snake bite or a bullet to the same vital organ will kill a rich person, or one with connections, as surely as it will the pauper or the nonentity. None of those extraneous factors will intrude as determinants of injury or death visited upon us by the laws of nature. Treatment available may depend on these extraneous factors, but that is determined by humans.

We know, and expect, that those theoretically extraneous factors will affect the outcome of laws administered by humans, and that they are not extraneous in practice.

But would we not expect that if evolution, one of the laws of nature, will change a population's skin tone upon one set of circumstances that the same circumstances would apply to all populations?

If the populations that settled Europe, beginning about 40,000 years ago were so closely looked after by a blind evolution, is there someone to whom the Lapps, the Inuit, the Eskimos, the Selkmans, the Far Eastern populations, or the Native Americans can appeal, should they be misguided enough to think they have been short changed? We overlook for the moment the fact that nature may have done the Caucasians no favor. The pale skin may or may not protect us against rickets and a host of diseases that come later in life. But it is true also that pale skin leaves us much more vulnerable to melanoma and other diseases in advanced age. Getting too much vitamin D can be dangerous and has been associated with kidney stones as well as other damage to the kidneys and the heart.

The evidence is almost entirely statistical, a tricky business at best. We really do not know who has been short changed. The tradeoff may have been for us a sorry bargain.

Underlying the entire evolutionary explanation is an unspoken assumption that evolution is a nursemaid, a nanny, who will jump in to correct any deficiency we may have as a protection against disease, or perhaps even discomfort. Think of all the diseases and debilities to which we are subject, and the inadequacy of our defenses against them. Why is there such an emergency response by nature to this particular threat, which, upon examination into the whole scheme of things, is relatively minor?

Scientists have figured out the answers to very serious problems, but their efforts to explain the uniquely pale skin of Caucasians have been a flop. They tell us it is an evolutionary response to insufficient sunlight. But not if you eat meat, fish, or fowl, though practically everyone does and always has. If most of the world's populations still have color, some tell us, it is because they eat animal flesh, hence do not need more vitamin D from sunshine. The combination that results in pale skin is, so we are told, less sunshine and more cereal; that is what made Caucasians so pale. But a number of large populations get plenty of cereals and as little or less sunshine than do the Europeans and their progeny, and they are not pale. Evolution can possibly explain lighter skin, tan, bronze, yellow, red, among others, but how does it explain colorless and pale from black or brown, especially when the results are so inconsistent?

I think that science has sold us a bill of goods. The body's response of lighter skin may indeed have merit. It can explain much, but not why Europeans are so uniquely pale. But that is not the only lemon we bought.

Evolution stumbles badly, even worse in my view, in explaining the vast distance between the human race and all other creatures on Earth. Science can tell us until hell boils over that there is only 2% or .2% or .02% difference genetically between us and the chimpanzee. Anyone with half an eye can see that there is a tremendous difference in body and behavior, no matter the genetics. They tell us there has been life on Earth for 3 ½ billion years and I know of no reason to doubt it.; human life for almost 4 million years, and no reason to doubt it; anatomically modern humans for 200,000, no reason to doubt it, but no big surprise if we were to learn that we were getting a boost from outside to help.

But 50,000 years and less, to communicate with sophisticated language, to create great art, great music, great literature, to figure out our physical place in the universe;; to build the cities we have built, to explore and measure the so tiny particles that make up the universe we live in, and the gigantic bodies that fill the great expanse of space that has been discovered; to create the electronic world we live in, or the quantum world we may soon harness for our own use and convenience? Was it impossible? Obviously I cannot say so, but to me it sounds unlikely—more so than the notion of hybridization, all evidence considered.

One of the anti-evolutionists, a gentleman named Humphrey Johnson, wrote in 1943 that there is a "greater difference between a man and a gorilla than between a gorilla and a daisy. One is as incapable as the other of creating a civilization."[278]

I can't say I consider that to be literally true, and there is possibly nothing else I could find in his writings that I could agree with at all. But with that one thought I can at least appreciate the gut feeling behind it. I believe we are creatures of evolution, but I believe that much of the evolution took place somewhere else, somewhere where the creatures have had a lot more time than have we in which to evolve.

I think we are beneficiaries of an evolution of intelligent creatures that has been at work far longer than the earthly humans have existed. I believe also that there is a lot going on out there of which we are totally unaware, and that there are forces in the universe of which our scientists have not yet dreamed. We have one race of ETs, the Greys, surreptitiously trying now to create a hybrid race with us, and there is no reason to dismiss the idea that someone else, some other race, has already done it in millennia past.

That, if true, and obviously I cannot know that for certain, still raises a lot of questions. How long has it been going on? That is the first question. That may not be the hardest question, but it must be first, as we are like the Nordics not just in skin tone. That is a resemblance that only applies to Caucasians. There are other similarities that apply to the whole human race. We, all of us, look like them. Two eyes, a nose, almost hairless bodies (compared to our cousins the chimpanzees) body style, arms and legs proportionately almost the same length as ours. What other extraterrestrials could pass themselves off, which some of the Nordic ETs apparently can do, at least temporarily, as natives of Earth.

Somewhere down the line there might possibly have been a mixture of genes, ours and theirs, long before there were anatomically modern humans, about 200,000 years ago. Perhaps it was the 300,000 deduced by Zecharia Sitchin. Perhaps it was before the first *Homo sapiens*, maybe Homo habilis, 3 ½ million years ago. If so, we are certainly seeing the results of *in vitro* fertilization.

Perhaps our species has been nursed along for the last 3 ½ million years, our evolution guided and accelerated, perhaps for our benefactor's own benefit. In time there were undoubtedly normal sexual relations. It cannot have escaped the readers' attention that contact by Earthlings with these ETs has been almost always a friendly one, leaving our contactees with a good, almost euphoric feeling. Seldom is that reported for contact with ETs other than Nordics. It is interesting also that some among these contactees have guessed that the relationship is based on something genetic. That doesn't make it true, but it is interesting nonetheless.

The second question that pops up is more difficult. It is one to which I have offered some guesswork in Chapter II. Though all humans have the bodies, and probably the minds of the ancient astronauts, why do only the Europeans and their descendants have their pale skin? My response is pure guesswork, but this entire subject is filled with guesswork, by hard scientists, and imaginative fiction writers alike. So I am not embarrassed to having tried my hand at it. A parent often favors the child that looks like him or her. The pale skin may have done it, at least initially.

There are other questions that raise their annoying heads, but they are of much less moment that those two. Obviously I cannot claim, and do not claim, that this tome has proved that the suggested scenario did, to a certainty, happen. But the mere possibility that it, or something like it, did happen, I find riveting for its possible significance.

Could it ever be proved or disproved? Theoretically it could. Recall from Chapter I the DNA of modern type that dates back in Africa to about two million years ago, the very beginning of Homo erectus, our very primitive ancestors. Dr. Akey acknowledges he has no clue as to where it came from. There is much evidence, convincing to some of us, that individual Nordics are working with our government, or a small supersecret segment of it. A comparison of genomes could probably tell us much. If so, such a comparison could be done only by the government. Don't hold your breath.

BIBLIOGRAPHY

This Bibliography contains listings only of those volumes and sources whose information is not sufficiently set forth in the body of the text or the end notes

Bahn, Paul G. and Jean Vertut. 1988. *Images of the Ice Age*. Leiscester, England: Windward.

Bickerton, Derek, 2009. *Adam's Tongue*. New York NY: Hill and Wang

Blum, Ralph and Judy Blum. 1974. *Beyond Earth*. New York: Bantam Books.

Brantingham, P. Jeffrey, ed., et al. 2004. *The Early Upper Paleolithic Beyond Western Europe*. Berkley: University of California Press.

Bryan, C.D.B. 1995. *Close Encounters of the Four h Kind*. London: Trafalgar Square.

Bury, J.B. 1902. A History of Greece to the Death of Alexander the Great. New York NY: Modern Library.

Carey, Nessa, 2012. New York: Columbia UP.

Carey, Thomas J. and Donald R. Schmitt. 2007. *Witness to Roswell*. Franklin Lakes, NJ: Career Press.

Cerve, Wishar S. 1997. Lemuria: The Lost Continent of the Pacific. San Jose, CA: Amorc Funds.

Chomsky, Noam. 1957. *Syntactic Structures*. The Hague: Mouton.

Cieza de León, Pedro de. 1864. *The Travels of Pedro de Cieza de León, Contained in the First Part of his Chronicle of Peru, AD 1532-1550.* London: Hakluyt Society.

Cody, F. and Bill, Buffalo. 2002. *Buffalo Bill's Life Story: An Autobiography,* New York: Turtle Point Press.

Cohane, John Philip. 1977. *Paradox: the Case for the Extraterrestrial Origin of Man.* New York: Crown.

Cooke, Patrick, *UFOs in History.* http://www.bibleufo.com/ufo.html.

Corso, Philip J. 1997. *The Day after Roswell.* New York: Pocket Books.

Cunnane, Steven C., ed., and Kathlyn Stewart, ed. 2010. *Human Brain Evolution: The Influence of Freshwater and Marine Food Resources.* Hoboken, NJ: Wiley-Blackwell.

Dolan, Richard M. 2002. *UFOs and the National Security State.* Charlottesville, VA: Hampton Roads Publishing.

Gardner, Helen. 1986. *Art Through the Ages* 8th ed. San Diego, CA: Harcourt, Brace. Jovanovich.

Gilfillan, Edward S. Jr. 1975. *Migration to the Stars.* New York: Robert B. Luce.

Good, Timothy. 1988. *Above top secret.* New York: Quill.

Good, Timothy. 2000. *Unearthly Disclosure.* London: Arrow Books.

Good, Timothy. 2007. *Need to know.* New York: Pegasus Books.

Greer, Steven M. 2001. *Disclosure.* Crozer, VA: Crossing Point.

Guilaine, Jean and Jean Zammit. 2005. *The Origins of War: Violence in Prehistory,.* Trans., M. Hersey. Oxford, UK: Blackwell Publishing.

Hadingham, Evan. 1979. *Secrets of the Ice Age.* New York NY: Walker.

Harrison, G.A. 1998. *Human Biology,* 3rd ed. Oxford: Oxford UP.

Hatzfeld, Jean. History of Ancient Greece. Trans. A.C. Harrison. New York, NY: W.W. Norton.

Hauser, Arnold. 1989. *The Social History of Art.* London: Routledge.

Hibbben, Frank C. 1958. *Prehistoric Man in Europe.* Norman: University of Oklahoma Press.

Hill, Paul R. 1995. *Unconventional Flying Objects.* Charlottesville, VA: Hampton Roads Publishing.

Hopkins, Budd and Carol Rainey. 2003. *Sight Unseen.* New York: Pocket Star Books.

Hopkins, Budd. 1987. *Intruders.* London: Sphere Limited.

Hopkins, Budd. 1996. *Witnessed.* New York: Simon and Schuster.

Horsley, Sir Peter. 1997. *Sounds From Another Room.* London: Leo Cooper.

Howe, Linda Moulton. 2001. *Glimpses of Other Realities: High strangeness, Vol. II.* Jamison, PA: LMH Productions.

Hurley, Matthew. 2003. *The Alien Chronicles.* Chester, United Kingdom: Quester Publications.

Hynek. Dr. J. Allen. *Night Siege*, 2nd ed., St. Paul, Minnesota: Llewellyn, 1998.

Jacobs, David M. 2000. *UFO & Abductions: Challenging the borders of knowledge.* Lawrence, KS: University Press of Kansas.

Kenneally, Christine. 2007. *The First Word.* Penguin Group: new York, NY.

Klein, Richard G. and Blake Edgar. 2002. *The Dawn of Human Culture.* New York: John Wiley and Sons.

Leroi-Gourhan, André. 1968. *The Art of Prehistoric Man in Western Europe.* London: Thames and Hudson.

Lorenzen, Coral, and Jim. 1969. *UFO: The Whole Story.* New York: Signet Books.

Lullies, Reinhard, and Max Hirmer. 1957. *Greek Sculpture.* New York NY: Harry N Abrams.

Lumsden, Charles J. and Edward O. Wilson. 1981. Genes, Mind and Culture. Cambridge: Harvard UP.

Mack, John E. 1999. *Passport to the Cosmos.* New York. Three Rivers Press.

Marshack, Alexander. 1991. *The Roots of Civilization.* Mt. Kisko, NY: Moyer Bell.

Moss, Chris. 2008. *Patagonia, a Cultural History.* New York: Oxford UP.

Osborne, Harold. ND. *South American Mythology.* London: Paul Hamlyn.

Owens, Francis, 1993. *The Germanic People.* New York: Barnes and Noble.

Pales, León and Marie Tassin De Saint Péreuse. 1976. *Les Gravures de la Marche: II Les Humains.* Paris: Orphys.

Pfeiffer, John E. 1982. *The Creative Explosion,* New York NY: Harper & Row.

Pinker, Steven. 1994. *The Language Instinct. New York, NY.* William Morrow.

Price, T.R., ed. *Europe's First Farmers.* 2000. Cambridge, United Kingdom: Press Syndicate of the University of Cambridge.

Randles, Jenny. 1994. Alien Contacts & Abductions: The Real Story From the Other Side. New York: Sterling.

Robins, Ashley H. 1991 *Biological perspectives on human pigmentation.* Cambridge University Press, pp. 195-208.

Salla, Michael E. 2004. Exopolitics: Political implications of the extraterrestrial presence. Tempe. AZ: Dandelion Books.

Sieveking, Ann. 1979. *The Cave Artists.* London: Thames and Hudson.

Sitchin, Zecharia, 1976. *The 12th Planet* New York: Harper.

Sitchin, Zecharia, 2010. Rochester. Vermont: Bear & Company.

Sparks, Jim, 2008, *The Keepers,* 2nd Rev. ed. Columbus, NC: Granite Publishing.

Tellier, Luc-Normand. 2009. *Urban world History: an economic and geographical perspective.* Quebec: Presses de l' Universite du Quebec.

Vallee, Jacques and Chris Aubeck. 2009. *Wonders in the sky.* New York NY: Jeremy Tarcher.

Vallee, Jacques. 1992. *UFO chronicles of the Soviet Union.* New York: Ballentine Books.

Wade. Nicholas. 2006. *Before the Dawn.* London: Penguin Books.

Ward, Paul von. 2004. *We've Never Been Alone*, Hampton Roads Publishing Company Inc.

Wells, Spencer. 2004. *The Journey of Man: A Genetic Odyssey.* New York: Random House.

Willcox, A.R. 1984. *The Rock Art of Africa.* London: Groom Helm.

Wolpoff, Milford H. 1989. "The place of Neandertals in human evolution," in *The Emergence of Modern Humans*, ed. Erik Trinkhaus. Cambridge: Cambridge UP.

ENDNOTES

1. https://www.earthfiles.com/news.php?ID=2016_and_category=Science
2. J.E. Pfeiffer, *The Creative Explosion*, 1.
3. K. K. Hirst, About.com Guide http://archaeology.about.com/od/upperpaleolithic/qt/Upper-Paleolitliic.htm
4. http://www.infoplease.com/ce6/society/A0860205.html
5. R. Klein and B. Edgar, *The Dawn of Human Culture*, 235.
6. N. Wade. *Before the Dawn, 9*, 12-13, 95.
7. N. Wade, http://www.nytimes.com/1999/12/07/science/genes-tell-new-story-on-the-spread-of-man.html?
8. R. Klein and B. Edgar, *The Dawn of Human Culture*, 181-185, 229.
9. https://www.cambridgedna.com/genealogy-dna-ancient-migrations-slideshow.php?view=step3
10. Ward, Paul von, *We've Never Been Alone,"* 35, 102.
11. E. Trinkaus, Schekman, R., ed. "European early modern humans and the fate of the Neandertals". *Proc. Natl. Acad. Sci. U.S.A.* (April 2004). **104** (18): 7367-72.
12. Ramachandran et al. (2005), "Support from the relationship of genetic and geographic distance in human populations for a serial founder effect originating in Africa". *Proc. Natl. Acad. Sci. U. S;* LJ, Handley et al

(September 2007); "Going the distance: human population genetics in a clinal world". *Trends in Genetics* **23** (9): 432-9.

13 http://www.nytimes.com/1999/12/07/science/genes-tell-new-story-on-the-spread-of-man.html?

14 http://www.vanughns-1-pagers.com/history/world-population-growth.htm

15 R. Klein and B. Edgar, *The Dawn of Human Culture,* 235.

16 R. Klein and B. Edgar, *The Dawn of Human Culture,* 187

17 N. Wade, *Before the Dawn,* 150, 151, 157.

18 J. Guilaine and J. Zammit. *The Origins of War: Violence in Prehistory,* 8. 2005. Trans., M. Hersey. Oxford UK: Blackwell Publishing

19 http://myfundi.co.za/e/War_in_Africa_I_Prehistory_to_1400#top-anchor

20 http://drbenkim.com/vitamin-d-facts.htm

21 https://www.ncbi.nlm.nih.gov/pubmed/20979596 Pigment Cell Melanoma Res. 2011 Feb;24(1): 136-47. doi: 10.1111/j.1755-148X.2010.00764.x. Epub 2010 Oct 6. The deceptive nature of UVA tanning versus the modest protective effects of UVB tanning on human skin. Miyamura Y, et al. PMID 20979596 http://www.skincancer.org/prevention/uva-and-uvb/landmark-research-links-melanoma-to-uv-radiation; published by *Skin Cancer Foundation.*

22 "Ancient DNA from the First European Farmers in 7500-Year-Old Neolithic Sites" *Science* 11 November 2005,Vol. 310 no. 5750 pp. 1016-1018.

23 Norton, H. L. et al (2006), "Genetic Evidence for the Convergent Evolution of Light Skin in Europeans and East Asians". *Molecular Biology and Evolution* **24** (3): 710-22.

24 Juzeniene, A. et al(2009). "Development of different human skin colors: A review highlighting photobiological and photobiophysical aspects". *Journal of Photochemistry ana Photobiology B: Biology* **96** (2): 93-100

25 http://mbe.oxfordjournals.org/content/24/3/710.abstract

26 Walters, KA; Roberts, MS (2008). "Dermatologic, cosmeceutic, and cosmetic development: Therapeutic and novel approaches." New York; *Informa Healthcare*

27 Relethford, J.H., (2000). "Human skin color diversity is highest in sub-Saharan African populations". *Human biology; an international record of research* **72** (5): 773-80.

28 Jablonski, N. G.; Jablonski, George; Chaplin (2000). "The evolution of human skin coloration". *Journal of human evolution* **39** (1): 57-106.

29 R. Blum and J. Blum. *Beyond Earth*, 39-40.

30 P. Cooke, *UFOs in History*, http://www.bibleufo.com/ufosl.htm

31 T. Good, *Above Top Secret*, 414.

32 E.S. Gilfillan, Migration to the Stars, 84.

33 Z. Sitchin, *The 12th Planet*, 171

34 Z. Sitchin, *There Were Giants Upon the Earth*, 198 et seq.

35 Z. Sitchin, *The 12th Planet*, 327.

36 Z. Sitchin, *The 12th Planet*, 343.

37 Z. Sitchin, *The 12th Planet*, 341.

38 Z. Sitchin, *The 12th Planet*, 348.

39 Z. Sitchin, *The 12th Planet*, 368-372.

40 http://graidi-taylor-rose101.com/different-types-of-extra-terrestrial-128060

41 *Chambers Dictionary of the Unexplained.* Ed. Una McGovern. Chambers, 2007. p. 489-490.

42 Clark, J. (2000). *Extraordinary Encounters: An Encyclopedia of Extraterrestrials and Otherworldly Beings.* ABC-CLIO. p. 187-188.

43 Bryan, C. D. B. (1995). *Close Encounters of the Fourth Kind: Alien Abduction, UFOs, and the Conference at M.I.T.* London: Trafalgar Square, pp. 30-31.

44 Kelley-Romano, Stephanie (2006). "Mythmaking in Alien Abduction Narratives". Published in *Extreme Deviance.* Ed. Erich Goode. Pine Forge Press, 2007. p. 51.

45   Bryan, C. D. B. (1995). *Close Encounters of the Fourth Kind: Alien Abduction, UFOs, and the Conference at M.I.T.* Knopf, p. 30-31.

46   Bryan, C. D. B. (1995). *Close Encounters of the Fourth Kind: Alien Abduction, UFOs, and the Conference at M.I.T.* Knopf, p. 30-31.

47   Clark, Jerome (2000). *Extraordinary Encounters: An Encyclopedia of Extraterrestrials and Otherworldly Beings.* ABL-CIO. p. 187-188.

48   Randles, Jenny (1994). *Alien Contacts & Abductions: The Real Story From the Other Side.* Sterling, p. 102-103.

49   M. Salla, **http:**//exopolitics.org/study-paper-8.htm

50   http://graidi-taylor-rose101.com/different-types-of-extra-terrestrial-128060

51   J.E. Mack, *Passport to the Cosmos*; D.M. Jacobs, *Secret Life;* D.M. Jacobs, *UFO ana Abductions:* B. Hopkins, *Intruders*; B. Hopkins, *Witnessed*; B. Hopkins and C. Rainey, *Sight Unseen:* B. Lamb and N. Lalich, *Alien Experiences.*

52   http//animalscience.ucdavis/edu/faculty/Anderson/research.htm; Environmental Health Perspectives, vol. 101, no. 4, September 1993, 208.

53   B. Hopkins, *Intruders,* 235.

54   http://www.hemmy.net/2006/06/19/top-10-hybrid-animals/

55   http://www.actionbioscience.org/biotech/margawati.html: E. T. Margawati, An ActionBioscience.org original article

56   B. Barcott. 2012. "Earth's Last Unexplored Wilderness is Your Living room" *Discover Magazine,* July/August, 35.

57   T. Good. *Unearthly Disclosure,* 137-138.

58   T. Good, *Unearthly Disclosure,* 132.

59   Horsley, Sir Peter, *Sounds From Another Room,* 180-196.

60   T. Good, *Alien Base,* 210-215

61   P. Pavlov et al, "Human presence in the European Arctic," *Nature* 413 (2001), 64-67.

62 M. Ho and P. Saunders, "The epigenetic approach to the evolution of organisms-with notes on its relevance to social and cultural evolution," 355.

63 E. Jablonka and M. Lamb, *Epigenetic Inheritance and Evolution: The Lamarckian Dimension*, 6, 26.

64 D. Dennett, "T Baldwin effect: A crane, not a skyhook."

65 C.J. Lumsden and E.O. Wilson, *Genes, Mind and Culture*, 295-301.

66 N. Carey, *The Epigenetics Evolution*, 18.

67 Waddington C. H. 1942. "Canalization of development and the inheritance of acquired characters."

68 D.N. Reznick et al, "Evaluation of the rate of evolution in natural populations of guppies (Poecillia)."

69 J.B. Losos et al, "Adaptive differentiation following experiment al island colonization in Anolis Lizards." *Nature* 387, May 1, 1997. pp. 70-72.

70 E. Watters, "DNA is not Destiny." *Discover*, 27 no. 11. November 2006. pp. 33-37.

71 N. Carey (2012). "Epigenetics in Action, Part 3," in *Natural History*, vol. 120. No 6, p 30, 3.1.

72 University of Maryland medical Center; *http://www.umm.edu/ency/article/000344.htm*: http://www.ncbi.nlm.nih.gov/pubmedhealth/PMH0001384/

73 http://www.mayoclinic.com/health/rickets/DS00813

74 Reference SNP (refSNP) Cluster Report: rs1426654 **clinically associated**. Ncbi.nhn.nih.gov (2008-12-30).

75 J.H.Relethford. Hemispheric difference in human skin color. Am J Phys Anthropol. 1997; 104:449-457. [PubMed]).

76 AC Ross, et al (2011). *Dietary Reference Intakes for Calcium and Vitamin D*. Washington, D.C: National Academies Press, p. 435.

77 R. Stokowski (2007). "A Genomewide Association Study of Skin Pigmentation in a South Asian Population". *The American Journal of Human Genetics* **81** (6): 1119-32; HapMap: SNP report for rs1042602, Hapmap.ncbi.nlm.nih.gov (2009-10-19); Shriver MD. Parra EJ, Dios

S, *et al.* (April 2003). "Skin pigmentation, biogeographical ancestry and admixture mapping". *Human genetics* **112** (4): 387-399.

78 HapMap: SNP report for rs642742. Hapmap.ncbi.nlm.nih.gov (2009-10-19).

79 Reference SNP (refSNP) Cluster Report: rs2424984Ncbi.nlm.nih.gov (2008-12-30); R. Valenzuela, R. K. Henderson et al. (2010). "Predicting Phenotype from Genotype: Normal Pigmentation". *Journal of Forensic Sciences* **55** (2): 315-22; HapMap: SNP report for rs2424984. Hapmap.ncbi.nlm.nih.gov (2009-10-19): Reference SNP (refSNP) Cluster Report: rs4911414 **clinically associated**. Ncbi.nlm.nih.gov (2008-12-30). Retrieved on 2011-02-27: Reference SNP (refSNP) Cluster Report: rs1015362 **clinically associated**. Ncbi.nlm.nih.gov (2008-12-30). N., Hongmei, (2009). "Genetic variants in pigmentation genes, pigmentary phenotypes, and risk of skin cancer in Caucasians". *International Journal of Cancer* **125** (4): 909-17; HapMap: SNP report for rs1015362. Hapmap.ncbi.nlm.nih.gov (2009-10-19).

80 M. Edwards, (2010). McVean, Gil. ed. "Association of the OCA2 Polymorphism His615Arg with Melanin Content in East Asian Populations: Further Evidence of Convergent Evolution of Skin Pigmentation". *PLoS Genetics* **6** (3): e1000867.

81 HapMap: SNP report for rs1800414. Hapmap.ncbi.nlm.nih.gov (2009-10-19).

82 Myles, et al. (2006), *Identifying genes underlying skin pigmentation differences among human populations;* Reference SNP (refSNP) Cluster Report: rs885479. Ncbi.nlm.nih.gov (2008-12-30): P. Shi, et al (2001). "Melanocortin-1 receptor gene variants in four Chinese ethnic populations". *Cell Research* **11** (1): 81-4: H.L. Norton, et al. (2006). "Genetic Evidence for the Convergent Evolution of Light Skin in Europeans and East Asians". *Molecular Biology and Evolution* **24** (3): 710-22.

83 http://stewartsynopsis.com/first_appearance_of_white_skin_i.htm

84 "European Skin Turned Pale Only Recently, Gene Suggests". A. Gibbons, (2007), AMERICAN ASSOCIATION OF PHYSICAL ANTHROPOLOGISTS MEETING: "European Skin Turned Pale Only Recently, Gene Suggests". *Science* **316** (5823): 364a.

85. http://www.webmd.com/diet/vitamin-ddeficiency

86. A.W. Yuen, N.G. Jablonski (January 2010). "Vitamin D: in the evolution of human skin colour". *Medical Hypotheses* 74 (1): 39-44.

87. http://drbenkim.com/dr-michael-holick-vitamin-d.htm

88. "Vitamin D". Mayo Clinic

89. http://www.differencebetween.net/science/difference-between-vitamin-d-and-vitamin-d3/

90. http://www.ehow.com/facts_7360459_difference-between-vitamin-d2-d3_.html

91. M.F. Holick et al. *Journal of Clinical Endocrinology and Metabolism*. March 2008; Endocrine Section. Department of Medicine, Boston University School of Medicine. Boston, Massachusetts 02118; "Vitamin D2 Is as Effective as Vitamin D3 in Maintaining Circulating Concentrations of 25-Hydroxyvitamin D."

92. http://www.ehow.com/facts_5535591_difference-between-vitamin-vitamin-d.html

93. The Misleading RDA for Vitamin D; http://naturalbias.com/vitamin-ds-flawed-recommended-daily-allowance/

94. http://www.raysahelian.com/vitamind.html

95. http://www.webmd.com/healthy-aging/news/20100426/higher-vitamin-d-better-golden-years, June 11, 2012

96. http://msdh.ms.gov/msdhsite/_static/230003124260003254.html

97. V. Willingham, November 30, 2010. CNN Medical http://www.cnn.com/2010/HEALTH/11/30/vitamin.d.calcium/index.html

98. http://www.mayoclinic.com/health/rickets/DS00813/DSECTION=risk-factors

99. http://www.naturalnews.com/003069.html#ixzzlfcqdBiTR

100. http://www.naturalnews.com/003069.html#ixzzlfcqdBiTR

101. VitaminDHealth.org

102 http://www.medwirenews.md/437/86415/Bone_Health/Rickets_mortality_html

103 http://www.naturalnews.com/003069.html#ixzzlfcqpyLFU

104 Aloia JF et al. (2010). "The 25(OH)D PTH threshold in black women". *The Journal of Clinical Endocrinology and Metabolism* **95** (11): 5069-73: Wang, TJ: et al. (2010). "Common genetic determinants of vitamin D insufficiency: a genome-wide association study". *Lancet* 376 (9736): 180-8.

105 J.S. Finkelstein et al. (2002). "Ethnic variation in bone density in premenopausal and early perimenopausal women: effects of anthropometric and lifestyle factors". *The Journal of clinical endocrinology and metabolism* 87 (7): 3057-67.

106 Harris, SS (2006). "Vitamin D and African Americans". *The Journal of nutrition* **136** (4): 1126-9; C. A. Gadegbeku, G. M. Chertow, (2009). "Cum Hoc, Ergo Propter Hoc: Health Disparities Real and Imagined". *Clinical Journal of the American Society of Nephrology* **4** (2): 251-3; J.F. Aloia, (2008); "African Americans, 25-hydroxyvitamin D, and osteoporosis: a paradox". *The American journal of clinical nutrition* 88 (2): 545S-550S.

107 I. Robertson et al, (1981), "The role of cereals in the aetiology of nutritional rickets: the lesson of the Irish National Nutrition Survey 1943-8". *The British journal of nutrition* **45** (1): 17-22; M. R. Clements, (1989). "The problem of rickets in UK Asians". *Journal of Human Nutrition and Dietetics* **2** (2): 105-116; Pettifor, J. (2004). "Nutritional rickets: deficiency of vitamin D, calcium, or both?". *The American journal of clinical nutrition* **80** (6 Suppl): 1725S-1729S.

108 GE Oramasionwu, et al (2008). "Adaptation of calcium absorption during treatment of nutritional rickets in Nigerian children". *The British journal of nutrition* 100 (2): 387-92; PR Fischer, et al (1999). "Nutritional rickets without vitamin D deficiency in Bangladesh". *Journal of tropical pediatrics* 45 (5): 291-3.

109 American Journal of Clinical Nutrition, Vol. 80. No. 6, 1725S-1729S, December 2004

110 G. Snellman, et al (2009). Z. Bochdanovits, ed. "Seasonal Genetic Influence on Serum 25-Hydroxyvitamin D Levels: A Twin Study". *PloS one* **4** (11): e7747; P. Lips, (2007). "Vitamin D status and nutrition in

Europe and Asia". *The Journal of steroid biochemistry ana molecular biology* **103** (3-5): 620-5.

[111] T. Hagenau et al, (2008). "Global vitamin D levels in relation to age, gender, skin pigmentation and latitude: an ecologic meta-regression analysis". *Osteoporosis International* **20** (1): 133-40.

[112] http://www.webmd.com/diet/vitamin-d-deficiency

[113] http://www.nap.edu/openbook.php?record_id=5776_and_page=250

[114] F.C. Hibben, *Prehistoric man in Europe,* 72, 77, 84, 95, 117, 149-153.

[115] S. C. Cunnane and K. Moore, *Human Brain Evolution: The Influence of Freshwater ana Marine Food Resources,* 126, 130.

[116] Klein. R. and B. Edgar, *The Dawn of Human Culture,* 238.

[117] D GITIG-October 28, 2011. "Europe's First Farmers Didn't Abandon Their Hunter-Gathering Habits." ARS Technica.

[118] F.C. Hibben, *Prehistoric man in Europe,* 95.

[119] F.C. Hibben, *Prehistoric man in Europe,* 117.

[120] F.C. Hibben, *Prehistoric man in Europe,* 152.

[121] http://mpkb.org/home/food/vitamind/dcontentfoods

[122] "Risks and Benefits" (PDF).

[123] "The Skin Cancer Foundation—The Vitamin D Dilemma | Vitamin D". Skincancer.org.

[124] A. Gibbons, sciencemag.org. https://www.freerepublic.com/focus/f-chat/1824685/posts

[125] R.E. Ackerman, 1996. "Bluefish Caves."

[126] C. Moss, *Patagonia, a Cultural History.*

[127] http://transientlives.blogspot.com/2009/03/ancient-people-of-tierra-del-fuego.html

[128] F.W. Sweet, December 15, 2002. *The Paleo-Etiology of Human Skin Tone.* National Institutes of Health.

129 http://www.math.hmc.edu/resources/odes/odearchitect/examples/el.pdf

130 http://www.bleaching-dental.com/articles/racial_characteristics_of_face.html

131 http://www.bleaching-dental.com/articles/racial_characteristics_of_face.html

132 "Ancient DNA from the First European Farmers in 7500-Year-Old Neolithic Sites" *Science* 11 November 2005, Vol. 310 no. 5750 pp. 1016-1018.

133 R. Selig A Quiet Revolution: "Origins of Agriculture in Eastern North America." *National Museum of Natural History Bulletin for Teachers*, Vol. 15, No. 2 1993. http://www.mesacc.edu/dept/d10/asb/lifeways/hg_ag/quiet_revolution.html

134 http://www.sweetwatermuseum.org/E_PrehPeople.htm

135 http://www.ou.edu/cas/archsur/flash/oklahoma.html

136 http://www.digitalhistory.uh.edu/database/article_display.cfm?HHID=660

137 http://www.britannica.com/EBchecked/topic/9647/origins-of-agriculture/10784/Early-history

138 http://www.britannica.com/EBchecked/topic/9647/origins-of-agriculture/10784/Early-history

139 G. W. Crawford" and M. Yoshizakib. "Ainu Ancestors and Prehistoric Asian Agriculture" *Journal of Archaeological Science* 1987,14, 201-213; Z. Jixu. 2006. "The Rise of Agricultural Civilization in China." *Sino Platonic Papers,* Number 175 December 2006. Philadelphia: University of Pennsylvania.

140 http://www.mongolia-attractions.com/agriculture-in-mongolia.html

141 D GITIG-October 28, 2011. "Europe's First Farmers Didn't Abandon Their Hunter-Gathering Habits." ARS Technica.

142 P. Balm and J. Vertut, *Images of the Ice Age,* 33.

143 E. Hadingham, *Secrets of the Ice Age,* 239; A. Leroy-Gourhan, *The Art of Prehistoric Man in Western Europe,* 100.

144 L. Pales and M. De Saint Péreuse, *Les Gravures de La Marche, II Les Humains.*

145 T. Dobzhansky, et al, (1963), "Two Views of Coon's "Origin of Races" with Comments by Coon and Replies". *Current Anthropology* **4** (4): 360-367.

146 http://blog.world-mysteries.com/science/how-many-major-races-are-there-in-the-world/

147 W.M. Bass, 1995. *Human Osteology: A Laboratory and Field Manual.* Columbia: Missouri Archaeological Society, Inc.; W.E. Eckert, 1997. *Introduction to Forensic Science.* CRC Press, Inc.; G.W. Gill, 1998. "Craniofacial Criteria in the Skeletal Attribution of Race." *Forensic Osteology: Advances in the Identification of Human Remains.* (2nd ed.) K. Reichs, (ed.), 293-315; W.M. Krogman, et al, 1986. *The Human Skeleton in Forensic Medicine,* Springfield: Charles C. Thomas.; "Racial Identification in the Skull and Teeth, Totem": *The University of Western Ontario Journal of Anthropology,* Volume 8, Issue 1 2000 Article 4.

148 http://www.bleaching-dental.com/articles/racial_characteristics_of_face.html

149 R.G. Klein and B. Edgar, *The Dawn of Human Culture,* 138.

150 R.G. Klein and B. Edgar, *The Dawn of Human Culture,* 125, 126.

151 K.D. Schick and N. Toth, *Making Silent Stones Speak* 263, 267, 283: R.G. Klein and B. Edgar, *The Dawn of Human Culture,* 142-144.

152 R.G. Klein and B. Edgar, *The Dawn of Human Culture,* 96-101.

153 C. Kenneally, *The First Word,* 212

154 R.G. Klein and B. Edgar, *The Dawn of Human Culture,* 170-186

155 A. Mitchell, "DNA Turning Story Into a Tell-All," *New York Times, Science times* 1, Jan 31, 2012.

156 O. Bar-Josef, Annu. Rev. Anthropol. 2002.31:363-393.

157 O. Bar-Josef, Annu. Rev. Anthropol. 2002.31:363-393, at 369.

158 R.G. Klein and B. Edgar, *The Dawn of Human Culture,* 238, 239.

159 J.F. Brantingham et al, ed. *The Early Upper Paleolithic Beyond Western Europe.*

160 J.E. Pfeiffer. *The Creative Explosion,* 1.

161. P.G. Bahn et al, *Images of the Ice Age*, 33.

162. A.R. Wilcox, *The Rock Art of Africa*, 160-161, 253, 237; P.G. Balm et al, *Images of the Ice Age*, 29-30.

163. A. Sieveking, *The Cave Artists*, 27.

164. G. Bazin, *A Concise History of Art*, 11.

165. A. Hauser, *The Social History of Art*, 2.

166. A. Leroi-Gourhan, *The Art of Prehistoric Man in Western Europe*, 503.

167. R.G. Klein and B. Edgar, *The Dawn of Human Culture*, 8.

168. R.G. Klein and B. Edgar, *The Dawn of Human Culture*, 187.

169. N. Wade, *Before the Dawn*, 96.

170. N. Wade. *Before the Dawn*, 97, 98.

171. A. Marshack, *Roots of Civilization*

172. http://www.donsmaps.com/cavepaintings2.html

173. http://news.bbc.co.uk/2hi/science/nature/975360.stm

174. S. Pinker, *The Language Instinct*, 353-354.

175. N. Chomsky. *Syntactic Structures*,

176. D. Bickerton, *Adam's Tongue*, 223-226.

177. D. Bickerton, *Adam's Tongue*, 39.

178. D. Bickerton, *Adam's Tongue*, 225-226.

179. D. Bickerton, *Adam's Tongue*, 210, 226.

180. D. Bickerton, *Adam's Tongue*, 232

181. E. Gans, "The Little Bang: The Early Origin of Language," Anthropoetics 5, no. 1 (Spring/Summer 1999).

182. C. Kenneally, *The First Word*, 212-214.

183. C. Kenneally, *The First Word*, 216.

184 C. Kenneally, *The First Word*, 221, 222.

185 http://www.huffingtonpost.com/2009/06/24/prehistoric-german-flute_n_220344.html: http://blogs.discovermagazine.com/80beats/2009/06/24/worlds-oldest-flute-shows-first-europeans-were-a-musical-bunch/

186 http://news.nationalgeographic.com/news/bigphotos/46910181.html

187 http://blogs.discovermagazine.com/80beats/2009/06/24/worlds-oldest-flute-shows-first- europeans-were-a-musical-bunch/

188 http://news.bbc.co.uk/2/hi/8117915.stm

189 M. Wolpoff, "Multiregional evolution" 89.

190 G.A. Harrison et al, *Human Biology*, 196

191 C.J Lumsden and E.O Wilson, *Genes, Mind and Culture*, 1, 295-301, 75-79.

192 C.J. Lumsden and O.E. Wilson, *Genes, Mind and Culture*, 295-301.

193 C.J. Lumsden and O.E. Wilson, *Genes, Mind and Culture*, 1.

194 N. Carey, *The Epigenetics Evolution* 9.

195 J. Hatzfeld, *History of Ancient Greece*, 10.

196 J.B. Bury, *A History of Greece to the Death of Alexander the Great*, 6.

197 J. Hatzfeld, *History of Ancient Greece*, 25-27.

198 H. Gardner, *Art Through the Ages*, 112, 113.

199 G. Bazin, *A Concise history of Art*, 67.

200 J.B. Bury, *A History of Greece to the Death of Alexander the Great*, 37.

201 R. Lullies and M. Hirmer, Plate 274, p XI.

202 F. Owen, 1993, *The Germanic People*. New York: Barnes and Noble.

203 115 http://www.ivy-rose.co.uk/References/Pupil

204 N. Wade, *Before the Dawn*, 8

205 T.J. Carey et al. *Witness to Roswell*, 141-144.

206 T. Good, *Need to Know*, 81.

207 C. and L. Lorenzen. *The Whole Story*, 66-68.

208 P.R. Hill. *Unconventional Flying Objects*, 245-247.

209 http://www.stargate-chronicles.com/home.html

210 http://www.stargate-chronicles.com/home.html

211 http://www.stargate-chronicles.com/site/et-observed-inside-space-shuttle-payload-bay/

212 G. LoBuono, *Exopolitics* vol. 3:2 (July 2009); http://exopoliticsjournal.com/vol-3/vol-3-2-LoBuono.htm

213 http://www.ufoevidence.org/documents/doc1156.htm

214 D.B. BrBPn, *Close Encounters of the Fourth Kind*, 334.

215 http://www.phils.com.au/col.wilson.htm

216 http://www.phils.com.au/col.wilson.htm

217 L.M. Howe, *Glimpses of Other Realities*, 212-216. 118 http://www.phils.com.au/col.wilson.htm

218 http://www.abovetopsecret.com/forum/thread60571/pgl

219 "Testimony of Captain Bill Uhouse," in S. M. Greer, ed., *Disclosure* 386-387.

220 http://www.phils.com.au/col.wilson.htm

221 T. Good, *Unearthly Disclosure*, 127.

222 T. Good, *Unearthly Disclosure*, 207-209: M. Salla, Eisenhower's 1954 Meeting with Extraterrestrials: The Fiftieth Anniversary of First Contact?, Research Study #8, 1/28/04; http://www.exopolitics.org/Study-paper-8.htm.; http://www.theblackvault.com/wiki/in-dex.php/Did_President_Eisenliower_Meet_with_Aliens_in_1954_(by_William_H._Moore).

223 http://research.borderlands.com/wiki/1954-04-16_-l letter_from_Gerald_Light_to_Meade_Layne

224 http://justgetthere.us/blog/archives/Hillary-Clinton-Desire-For-UFO-Extraterrestrial-Diplomacy.html

225 T. Good, *Need to Know,* 207-209.

226 T. Good, *Need to Know,* 208.

227 D. Phillips, "Testimony of Don Phillips," in S. M. Greer, ed., *Disclosure* 375-383, at 379

228 Personal notes from William Hamilton from a 1991 interview with Commander Suggs' son, Sgt. Charles Suggs, Jr.

229 J.P Cohane, *Paradox: The Case for the Extraterritorial Origin of man,* 154.

230 http://www.geocities.com/Area51/Shadowlands/6583/maji007.html

231 http://www.ufoevidence.org/documents/doc1856.htm

232 M. Salla, Eisenhower's 1954 Meeting with Extraterrestrials: The Fiftieth Anniversary of First Contact?, Research Study #8, 1/28/04; http://www.exopolitics.org/Study-paper-8.htm

233 T. Good, *Need to Know,* 81.

234 http://graidi-ttaylor-rose.suite101.com//different-types-of-extra-terrestrial-a128060

235 D. Worley, "The Beautiful Blondes and their Incredible Flying Machines," *UFO Universe,* Burlington UFO Center, Home Page, Winter 1997 issue; http://www.burlingtonnews.net/nordics.html

236 R.L. Thompson, *Alien Identities—Ancient Insights into Modern UFO Phenomena*; http://www.livinginthelightms.com/shambhalaufos

237 P.R. Hill, *Unconventional Flying Objects,* 247-249

238 M. Hurley, *The Alien Chronicles,* 122.

239 T. Good, *Alien Base,* 29-44.

240 T. Good, *Alien Base,* 38.

241 Cervé, Wishar S., *Lemuria: The Lost Continent of the Pacific,* 250-252.

242 C.L. Brace (1996). A.M. Haeussler, et al, eds. "Cro-Magnon and Qafzeh— vive la Difference" (PDF). *Dental anthropology newsletter: a publication of the Dental Anthropology Association* (Tempe, AZ: Laboratory of Dental

Anthropology. Department of Anthropology, Arizona State University) **10** (3): 2-9.

243 http://www.blavatsky.net/newsletters/cro_magnon.htm

244 Trinkaus, Erik (April 2004). Schekman, Randy, ed. "European early modern humans and the fate of the Neandertals". *Proc. Natl. Acad. Sci. U.S.A.* **104** (18): 7367-72.

245 Fagan, B.M. (1996). *The Oxford Companion to Archaeology.* Oxford, UK: Oxford University Press, 864.

246 T. Good, *Alien Base,* 37.

247 HudsonAlpha, Institute for Biotechnology Edition: Fall 2009; http://www.hudsonalpha.org/education/outreach/basics/eye-color

248 https://www.23andme.com/health/Hair-Color/

249 L Raffensperger, 3 May 2012, http://www.newscientist.com/article/dn21779-blonde-hair-evolved-independently-in-pacific-islands.html

250 A.H. Robins,. *Biological perspectives on human pigmentation.* Cambridge University Press, 1991. pp. 195-208.

251 L.L.Cavalli-Sforza, et al, 1994. *"Europe". The History and Geography of Human Genes.* Princeton, New Jersey: Princeton University Press, p. 266. 29.

252 C. Keyser et al. 2009. "Ancient DNA provides new insights into the history of south Siberian Kurgan people" in *Human Genetics.* Vol. 126, Number 3, 395-410.

253 D.L. Duffy et al. 2004. "Interactive effects of MC1R and OCA2 on melanoma risk phenotypes" in *Human Molecular Genetics* 13:447-461.; P. Frost, 2006. "European hair and eye color—A case of frequency-dependent sexual selection?" in *Evolution and Human Behavior* 27:85-10.

254 B.K. Rana, B.K., 1999. High polymorphism at the human melanocortin 1 receptor locus. *Genetics* 151:1547-1557.

255 Harding, R.M. et al,. 2000. "Evidence for variable selective pressures at MC1R" *American Journal of Human Genetics* 66:1351-1361; A.R. Templeton,. 2002. "Out of Africa again and again." *Nature* 416:45-51.

[256] Frost, P. 2006. "European hair and eye color—A case of frequency-dependent sexual selection?" *Evolution and Human Behavior* 27:85-103; K. Makova, K, and H. Norton, 2005. "Worldwide polymorphism at the MC1R locus and normal pigmentation variation in humans". *Peptides* 26:1901-1908.

[257] K. Briggs, *An Encyclopedia of Fairies, Hobgoblins, Brownies, Boogies, and Other Supernatural Creatures,* "Golden Hair", p194.

[258] T. Good, *Alien Base,* 41, 43

[259] T. Good, *Alien Base,* 50-54

[260] T. Good, *Alien Base,* 54-55

[261] T. Good, *Alien Base,* 87-98

[262] T. Good, *Alien Base,* 166-168

[263] T. Good, *Alien Base,* 168-169

[264] T. Good, *Alien Base,* 204-205

[265] T. Good, *Alien Base,* 223, 227.

[266] T. Good, *Alien Base,* 240-248.

[267] T. Good, *Alien Base,* 400.

[268] http://www.nwcreation.net/noahlegends.html

[269] http://www.nwcreation.net/noahlegends.html

[270] H. Osborne, ND, *South American Mythology,* 72, citing *Cieza de León, Part I, Chapter 100.*

[271] H. Osborne, ND, *South American Mythology,* 74, citing *Cieza de León, Part II, Chapters 4 and 5.*

[272] H. Osborne, ND, *South American Mythology,* 87, citing Antonio de la Calancha, 1638, *Crónica moralizada del orden de San Agustin en el Perú.*

[273] H. Osborne, ND, *South American Mythology,* 107.

[274] H. Osborne, ND, *South American Mythology,* 81, 82.

275 Business News, May 21, 2012. "Paralyzed patients control robotic arm using brain activity."; http://www.healio.com/orthotics-prosthetics/prosthetics/news/online/%7BB6241AC2-3FD3-4A67-91D7-32DA792BE29C%7D/Paralyzed-patients-control-robotic-arm-using-brain-activity

276 http://keloki-biblebasics.blogspot.com/2006/10/nephalim-ufos-and-alien-encounters.html

277

278

www.ingramcontent.com/pod-product-compliance
Lightning Source LLC
Chambersburg PA
CBHW060351080526
44583CB00012B/263